THE Energy REPORT

VOLUME 1

Competition, Competitiveness, and Sustainability

Department of Trade and Industry

April 1995

HMSO

© *Crown copyright 1995*
Applications for reproduction should be made to HMSO

ISBN 0 11 515362 4

British Library Cataloguing-in-Publication Data.
A catalogue record for this book is available from the British Library.

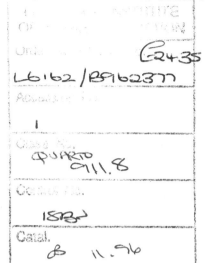

Published by HMSO and available from:

HMSO Publications Centre
(Mail, fax and telephone orders only)
PO Box 276, London SW8 5DT
Telephone orders 0171 873 9090
General enquiries 0171 873 0011
(queuing system in operation for both numbers)
Fax orders 0171 873 8200

HMSO Bookshops
49 High Holborn, London WC1V 6HB
(counter service only)
0171 873 0011 Fax 0171 831 1326
68-69 Bull Street, Birmingham B4 6AD
0121 236 9696 Fax 0121 236 9699
33 Wine Street, Bristol BS1 2BQ
0117 9264306 Fax 0117 929 4515
9-21 Princess Street, Manchester M60 8AS
0161 834 7201 Fax 0161 833 0634
16 Arthur Street, Belfast BT1 4GD
01232 238451 Fax 01232 235 401
71 Lothian Road, Edinburgh EH3 9AZ
0131 228 4181 Fax 0131 229 2734
The HMSO Oriel Bookshop
The Friary, Cardiff CF1 4AA
01222 395548 Fax 01222 384347

HMSO's Accredited Agents
(see Yellow Pages)

and through good booksellers

Standing order service

Placing a standing order with HMSO BOOKS enables a customer to receive future editions of this title automatically as published.

This saves the time, trouble and expense of placing individual orders and avoids the problem of knowing when to do so.

For details please write to HMSO BOOKS, Standing Orders Dept. Publications Centre, PO Box 276, London SW8 5DT and quote reference 12.01.018.

The standing order service also enables customers to receive automatically as published all material of their choice which additionally saves extensive catalogue research. The scope and selectivity of the service has been extended by new techniques, and there are more than 4,000 classifications to choose from. A special leaflet describing the service in detail may be obtained on request.

Designed and typeset by
Rye Associates, London

 Printed on Recycled Paper.

Contents

Chapter		Paragraph	Page
	MINISTERIAL FOREWORD		viii

Part One: The Main Themes

1	**INTRODUCTION**		1
2	**COMPETITION IN ENERGY MARKETS**		
	COMPETITION IN ELECTRICITY MARKETS	2.5	6
	Generation		
	Supply		
	COMPETITION IN GAS MARKETS	2.18	11
	ELECTRICITY AND GAS PRICES	2.22	12
	MARKETS IN TRANSITION	2.38	17
3	**ENERGY COSTS, ENERGY EFFICIENCY, AND INDUSTRIAL COMPETITIVENESS**		
	ENERGY AS A COST	3.3	25
	UK ENERGY PRICES COMPARED WITH OTHER COUNTRIES	3.6	25
	ENERGY EFFICIENCY	3.13	29
	Trends		
	The Government's approach		
	Competitiveness		
	Public and domestic sectors		
4	**FUTURE UNCERTAINTIES**		
	GLOBAL ENERGY FUTURES	4.3	40
	UK ENERGY PROJECTIONS	4.14	44
	TECHNOLOGICAL CHANGE	4.26	47
	CHANGES IN DEMAND	4.31	48

Chapter		Paragraph	Page
5	**ENERGY AND THE ENVIRONMENT**		
	SUSTAINABLE DEVELOPMENT	5.1	52
	THE LINK TO COMPETITIVENESS	5.4	52
	AIR QUALITY	5.12	55
	CLIMATE CHANGE	5.16	57
6	**INTERNATIONAL ENERGY AND ENVIRONMENT INITIATIVES**		
	EUROPEAN UNION	6.2	62
	Energy Green Paper		62
	Trans European Networks		62
	Liberalisation and the single purchaser model		63
	Carbon/energy tax		64
	Large Combustion Plant Directive		64
	Energy technology programmes		64
	Energy efficiency programmes		65
	Directive on integrated resource planning		65
	Voltage harmonisation		65
	INTERNATIONAL	6.23	66
	Energy Charter Treaty		66
	International Energy Agency		67
	United Nations		67
	CONVENTION ON NUCLEAR SAFETY	6.35	69
	ACTION PLAN FOR UKRAINE	6.36	69

Chapter		Paragraph	Page

Part Two: The Energy Industries

7 THE ENERGY INDUSTRIES

	Paragraph	Page
ENERGY COMPANIES IN INTERNATIONAL MARKETS	7.2	74
ENERGY CAPITAL GOODS	7.8	75
OIL AND GAS SUPPLIES	7.11	76
ENERGY EFFICIENCY SUPPLIES	7.15	77

8 COAL

	Paragraph	Page
STRUCTURE OF THE INDUSTRY IN 1994	8.1	80
INDUSTRIAL PRICES	8.6	82
PRIVATISATION	8.7	83
Sale of British Coal's mining business		83
The Coal Authority		84
Non-operational property		85
COAL MINING SUBSIDENCE	8.16	86

9 ELECTRICITY

	Paragraph	Page
GENERATION IN THE UK	9.4	88
GENERATION IN ENGLAND AND WALES	9.7	89
TRANSMISSION IN ENGLAND AND WALES	9.20	94
DISTRIBUTION IN ENGLAND AND WALES	9.25	95
SUPPLY IN ENGLAND AND WALES	9.29	95
THE ELECTRICITY MARKET IN SCOTLAND	9.37	98
THE ELECTRICITY MARKET IN NORTHERN IRELAND	9.41	99
NUCLEAR ELECTRICITY	9.46	100
NEW AND RENEWABLE SOURCES OF ELECTRICITY	9.55	102
AUTOGENERATORS AND CHP	9.62	105

Chapter	Paragraph	Page

10 GAS SUPPLY

	Paragraph	Page
UK MARKET DEVELOPMENTS	10.2	108
STRUCTURE OF THE MARKET	10.6	109
THE GAS BILL	10.12	111
GAS PRICES	10.18	112
EXTERNAL TRADE IN GAS	10.22	113

11 OIL AND GAS PRODUCTION

	Paragraph	Page
UK PRODUCTION AND RESERVES	11.1	118
LICENSING	11.7	120
ENVIRONMENTAL MATTERS	11.10	121

12 DOWNSTREAM OIL

	Paragraph	Page
DEVELOPMENTS IN UK MARKETS	12.1	124
TRANSPORT FUELS	12.3	125
Petrol		
DERV		
Transport fuel retailing		
Jet fuel		
OTHER FUELS	12.9	127
IMPLICATIONS FOR REFINERS	12.11	128
PRODUCT EXPORTS	12.13	128
THE UK CRUDE SLATE	12.14	128
THE UK INDUSTRY	12.15	129
Price trends		130
THE ENVIRONMENTAL CHALLENGE	12.22	132

Page

Annex

SUMMARY OF MAIN EVENTS IN 1995/96 134

Appendices

1	STATISTICS – PRIMARY ENERGY PRODUCTION AND CONSUMPTION	136
	– IMPORTS AND EXPORTS	141
	– FINAL ENERGY CONSUMPTION	142
	– PRICES	147
	– INTERNATIONAL COMPARISONS	152
	– TECHNICAL NOTES	154
	– STATISTICAL TABLES	157
2	ENERGY-RELATED ATMOSPHERIC EMISSIONS	174
3	POWER STATIONS, OIL REFINERIES, COAL MINES	184
4	DECOMMISSIONING AND RADIOACTIVE WASTE ISSUES	192
5	PUBLICATIONS	194
6	GLOSSARY OF TERMS	206
7	THE ENERGY ADVISORY PANEL: LIST OF MEMBERS	208

INDEX 209

FOREWORD BY TIM EGGAR, MINISTER FOR INDUSTRY AND ENERGY

This is the second Annual Energy Report. It is in two parts, published separately: this Volume, Competition, Competitiveness, and Sustainability, and Volume 2, Oil and Gas Resources of the United Kingdom ("The Brown Book"). The preparation of the Report has been assisted by the Energy Advisory Panel, though its contents remain the responsibility of the Government. I am most grateful to the Panel for its help.

The main themes of Volume 1 are the competitiveness of the UK economy, and how the Government's market-based policies are helping to enhance our prospects for the future, both as a nation and as individuals; and the growing recognition that we will have to have a greater awareness in future of the environmental aspects of energy production and use.

The apparently disparate objectives of national competitiveness and sustainable development do ultimately have a common theme - they are both essentially about making the best possible use of resources. The Government's view is that the optimal use of resources can for the most part be left to competitive forces in free markets. Competition has many benefits - it gives consumers choice, it maximises the pressure to provide what people actually want, rather than what the supplier is prepared to offer, and it gives companies an incentive to run themselves efficiently. The gas and electricity markets in the UK are not yet fully free, but we are already seeing some of the benefits of the steps taken so far - lower prices for consumers, higher productivity among suppliers, and the prospect of a wider range of service. These issues, amongst others, are considered within this Report.

Looking to the longer term, free markets will provide a better means of providing secure and diverse supplies of energy than would have been the case under the old nationalised energy industries. There is already more diversity of primary fuel for electricity supply than we have ever had. And we now have a range of companies which depend for their commercial survival on being able to meet our energy needs ten or twenty years hence, rather than having to depend on a single nationalised monolith.

The year 1994/95 has been an active one in many parts of the energy industry. Electricity and gas, in particular, have found themselves in the news. Some of the events have related to the workings of private companies, in which it is not appropriate for the Government to interfere. Other activities, which are reported in this Volume, relate to the Government, either domestically or internationally, or to the industry regulators. Some of these events are still unfolding, so the story may be incomplete. But what has become transparently clear over the last year is the genuine independence of the gas and electricity regulators, whose offices were established to operate without interference from Government.

The Government's role is to stay well away from decisions of detail - the market is perfectly capable of sorting itself out, and in a way which maximises effective use of resources. But when it comes to the environment, the market cannot always make the right choices, since many of the ways in which energy production and use affect the environment are outside the normal processes of exchange - so-called "externalities". It is right, therefore, that governments, both in the UK and elsewhere, should set the legal and regulatory framework within which markets can operate - but should then leave them to get on with it. This is how it is happening in the UK; and we are leading the world in opening our energy markets in this way.

Privatisation and liberalisation have significantly increased the number of players in energy markets. As a result, the debate over choices is more open, and, I believe, much better informed. The Government has a part to play by telling people about developments in energy markets from its own perspective. This was why an annual report of this kind was started. I believe it is meeting its objectives. Whether or not the reader agrees with every detail, the Energy Report provides a wealth of information and comment which comprehensively illuminates the UK energy scene.

Part One:
The Main Themes

CHAPTER 1

INTRODUCTION

1 INTRODUCTION

1.1 This is the Government's second annual Energy Report. Its purpose is to help competitive markets to develop by setting out the key elements of energy policy and the main driving forces, both external and internal. In preparing its Report, the Government has been advised by an independent panel of energy experts (whose membership is given in Appendix 7). This Energy Advisory Panel has identified the main elements which it feels should be covered in the Report, and has offered suggestions on the structure: but it is the Government's Report, and the Panel is not necessarily committed to all the details.

1.2 The main themes of the report are *competitiveness,* and the contribution the Government's market-based energy policies make to the improvement of the UK's relative economic position; and *sustainable development,* and the growing recognition of the potential scale of the environmental impacts associated with energy use. The ways in which energy is produced and used are central to the achievement of both policies. The two policies lead in the same direction: in essence, both are about making the best use of resources.

1.3 A further theme is the *growth of competition in energy markets*. Competition and competitiveness are separate concepts, though they are linked. The Government's approach to energy policy is that, so far as possible, decisions should be left to markets operating in a competitive environment. Important steps are being taken to increase the amount of competition: the years from now until 1998 are stages in a transitional process leading towards the full unwinding of the franchise market for the sale of electricity and gas. After 1998, all consumers – domestic, commercial, and industrial – will have the opportunity to choose their energy supplier. The events of 1998 can already be seen on the horizon. It is important for all participants in energy markets to understand what is likely to happen then, and what is happening in preparation. The events of 1994/95 reported here should be viewed in this context.

1.4 As competition develops, consumers should see the benefit both of lower prices, as a result of greater efficiency, and of better services. The process of transition is still in progress, but a number of benefits have already been achieved as a result of the Government's policies. Chapter 2 reports on progress.

1.5 Energy can be seen as one of the 'currencies' of a modern industrial society, and the energy industries can make a major contribution both to wealth creation and to the general quality of life. The way in which improved performance in the energy industries may be expected to flow through into UK competitiveness is a further theme of the Report. Chapter 3 looks at UK energy costs, relative to those of competitors, and considers how these have changed in recent years.
The contribution of the energy industries themselves to the UK's international

trading performance – both in sales of primary energy, notably oil and gas, and in sales of professional and financial services, drawing on the experience of the newly-privatised companies – is reviewed briefly in chapter 7.

1.6 The wider environmental impacts of energy use will determine whether any given use of energy is consistent with sustainable development. Chapter 5 considers the environmental impacts associated with the use of energy and, in particular, the potential impact of CO_2 emissions on the global climate. As a basis for this discussion, chapter 4 gives a broad assessment of future prospects in energy markets, both in the UK and in the world as a whole.

THE AIMS OF UK ENERGY POLICY

The aim of the Government's energy policy is **to ensure secure, diverse, and sustainable supplies of energy in the forms that people and businesses want, and at competitive prices.** The Government believes that this aim will best be achieved by means of competitive energy markets working within a stable framework of law and regulation to protect health, safety, and the environment. Government policies also aim to encourage consumers to meet their needs with less energy input, through improved energy efficiency.

The **key elements of** the policy are:

- to encourage competition among producers and choice for consumers, and to establish a legal and regulatory framework to enable markets to work well;

- to ensure that service is provided to customers in a commercial environment in which customers pay the full cost of the energy resources they consume;

- to ensure that the discipline of the capital markets is applied to state-owned industries by privatising them where possible;

- to monitor and improve the performance of the remaining state-owned industries, while minimising distortion;

- to have regard to the impact of the energy sector on the environment, including taking measures to meet international commitments;

- to promote energy efficiency;

- to safeguard health and safety;

- to promote wider share ownership.

ENERGY STATISTICS

Much of the data presented in this issue of the Energy Report has been obtained from the Department of Trade and Industry's statistical information systems, which hold data from a wide variety of sources. Figures for 1994 are provisional and subject to revision.

A comprehensive statistical account of the UK energy industry is provided by the Department's annual *Digest of United Kingdom Energy Statistics*, which consists of tables, charts, and commentary describing the production, use of, and trade in fuels over the five years 1990 to 1994. Additional sections cover fuel prices, long term price trends, renewable sources of energy, and combined heat and power. The 1995 edition of the Digest will be published by HMSO on 27 July 1995. More details can be obtained from the address below.

Up-to-date energy statistics can be found in the DTI's monthly publication *Energy Trends*. This is available on annual subscription only: for further details or a sample copy, contact:

**Mike Ward, EPA 4a, DTI, Room 3.3.15,
1 Palace Street, London SW1E 5HE
TEL: 0171 238 3576**

CHAPTER 2

COMPETITION IN ENERGY MARKETS

2 Competition in Energy Markets

2.1 The privatisation and liberalisation of the energy industries lie at the heart of the Government's energy policy. These industries are, so far as possible, to be like all others, where privately-owned companies operate in a competitive environment. The last annual Energy Report gave details of the new structures evolving in the electricity and gas industries. This chapter brings the story up to date.

2.2 Privatisation and liberalisation are distinct. Privatisation is the transfer of ownership from the public to the private sector. As a result, business objectives are clarified, with a profit motive providing an incentive to greater efficiency; managers become accountable to shareholders, who judge their performance against that of other companies; and Government interference, which has so often in the past undermined both managerial responsibilities and incentives to cut costs, is removed.

2.3 Liberalisation is the introduction of competition into markets. It increases incentives to minimise costs, to produce the quality and range of goods and services demanded by customers, to innovate, and to price in line with costs. Above all, competition gives consumers the freedom to choose which company will supply them.

2.4 This chapter is in two parts. The first shows how the structure of the energy markets, expressed in terms of market shares, has been changing over the past few years as new entrants have arrived. It also looks at the accompanying changes in prices. But looking only at market shares does not tell the whole story: the degree of competition also depends on the existence of barriers to entry, and on the behaviour of the various players. The second part of the chapter, *Markets in Transition*, looks at possible developments in the energy industry in the years to 1998 and beyond.

COMPETITION IN ELECTRICITY MARKETS

2.5 The production and supply of electricity are now carried out by a range of different companies. The market for generation is liberalised; the market for supply is partly liberalised. Transmission (long distance transport and control of the system) and distribution (delivery to customers through local networks) remain monopoly activities of the National Grid Company (NGC) and the Regional Electricity Companies (RECs) respectively in England and Wales, and of the two vertically-integrated Public Electricity Suppliers (PESs) in Scotland. A more detailed description of the electricity industry is given in chapter 9.

Generation

2.6 Generation was formerly the responsibility of the Central Electricity Generating Board (CEGB) in England and Wales, the North of Scotland Hydro-Electric Board and the South of Scotland Electricity Board in Scotland, and Northern Ireland Electricity in Northern Ireland. About 250 other companies produced some electricity for their own use (autogeneration) or as a by-product of local combined heat and power (CHP) schemes. Today, generation in the UK is carried out by: 12 companies (including 2 nuclear companies) which were formerly part of the old nationalised industries; some 10 major new entrants who have come into the market since 1989; and approximately the same number of autogenerators as before. Renewable energy sources are discussed in chapter 9.

2.7 Market shares can be considered in terms of shares of total output, shares of capacity, and shares of baseload and non-baseload output. Chart 2.1 shows the changes in market shares of total output in the UK. It shows the fall in the share of the privatised generators and the growth in the contribution of nuclear sources and of the new entrants. Most of the changes in market share have occurred over the last two years.

Chart 2.1

The market shares in generation, UK, 1989 and 1994

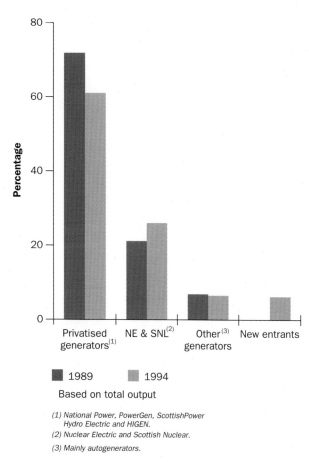

Based on total output

(1) National Power, PowerGen, ScottishPower Hydro Electric and HIGEN.
(2) Nuclear Electric and Scottish Nuclear.
(3) Mainly autogenerators.

Source: Department of Trade and Industry

2.8 In **England and Wales,** National Power has experienced the largest fall in market share during this period, particularly in 1993/94; PowerGen has seen a smaller decline. Nuclear Electric (NE) and the new entrants have been the beneficiaries, with NE's increase in share due to its improved performance from its existing assets. Nevertheless, the combined share of National Power and PowerGen remains above 50 per cent. chapter 9 deals with generation in Scotland and Northern Ireland.

2.9 At vesting in 1990, the generators' shares of output and of capacity were very similar. The latest statistics show that National Power and PowerGen have a greater share of capacity than of output, whereas the reverse is true for NE. National Power's and PowerGen's greater access to spare capacity gives them more operational and commercial flexibility than the other companies. Chart 2.2 shows the generators' share of new capacity, i.e. that opened since vesting and that under construction. Almost half of the new capacity is owned by new entrants, with National Power and PowerGen together accounting for about 39%. NE's new capacity is Sizewell B.

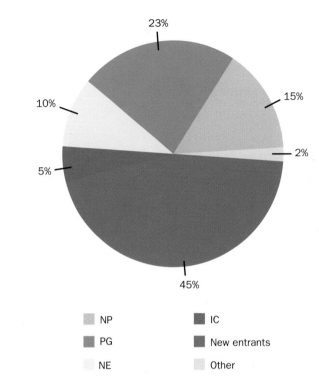

Chart 2.2

The generators' share of new capacity, England and Wales[1]

- NP
- PG
- NE
- IC
- New entrants
- Other

(1) Transmission contracted generation plant registered with the NGC since vesting and up to February 1995 (excluding plant not under construction).
NP=National Power, PG=PowerGen, NE=Nuclear Electric, In=Interconnectors and pumped storage, Other are mainly British Nuclear Fuels Ltd and pooled renewables.

Source: Department of Trade and Industry

2.10 Most of the capacity owned by the new entrants is intended to run on baseload (throughout the day and night). National Power and PowerGen continue to dominate the non-baseload market (extra daytime output), where they face little competition in setting the Pool price (see chapter 9). The Office of Electricity Regulation (OFFER) has estimated that the plant owned by these two companies effectively set the Pool price for over 90% of the time between June 1991 and January 1994.

2.11 In the near future, the level of competition is likely to be affected by any sale or disposal of plant by National Power and PowerGen as a result of their undertakings to the Director General of Electricity Supply (DGES) (see paragraph 9.12). The sale of 6 GW of plant would, in total, account for about 10% of total capacity available to the system in England and Wales, but would still leave these two generators with about 55% of total capacity. However, the presence of new owners should have an impact on the non-baseload Pool price.

Supply

2.12 At vesting, competition was introduced into the supply of electricity for customers using large amounts of electricity (over 1 MW). In April 1994, competition for medium-sized users (over 100 kW) began, and on 1 April 1998, all consumers of electricity will be able to choose their electricity supplier. Different arrangements apply in Northern Ireland. In England and Wales, there are over 5,000 large users, accounting for about 30% of the market in consumption terms; over 45,000 medium-sized users accounting for about 15% of the market; and 22 million (mainly domestic) customers accounting for the remaining 55%. Thus, as from April 1994, just under half of the market in England and Wales was open to competition.

2.13 The number of companies supplying electricity as their primary business has more than doubled since privatisation. This is shown in Chart 2.3.

Chart 2.3

Number of companies supplying electricity, UK, 1989 to 1994

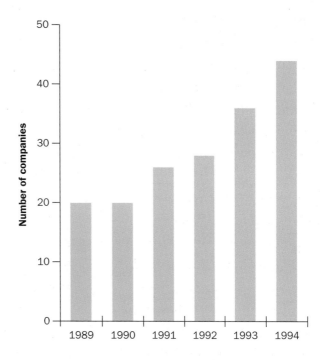

Includes Major Power Producers (i.e. those whose primary business is the generation of electricity), RECs and other pool members

Source: Department of Trade and Industry

2.14 In addition, there are the 250 or so autogenerators.

OFFER monitors market shares by an annual survey of supply to the non-franchise sector in England and Wales. Chart 2.4 shows the suppliers' changing shares of the above 1 MW sector in terms of output supplied. The survey also covers the number of sites supplied. More detailed figures are in Table 9.2.

2.15 Looking at the sector in terms of the number of sites supplied shows a similar trend as for output, i.e. losses for the first tier RECs and gains for second tier suppliers. However, the "Others" share has been broadly constant, which implies that the generators (which account for most of the Others) have concentrated their efforts on the sites with the largest consumption.

Chart 2.4
The market shares in the above 1 MW sector, England and Wales, 1990/91 and 1994/95

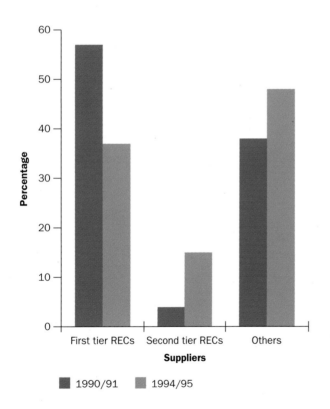

Based on output supplied.
Figures for 1994/95 are based on company estimates made in July 1994.
First tier RECs are those operating within their own supply area. Second tier RECs are those supplying outside their own supply areas.
Others include National Power, PowerGen, Hydro-Electric, ScottishPower, Nuclear Electric and Independent suppliers.

Source: OFFER

COMPETITION IN ENERGY MARKETS

2.16 For the 100 kW to 1 MW sector in 1994/95, the first tier RECs accounted for about 75% of supply, and the second tier RECs for 19%, in terms of number of sites. The shares of output are broadly similar.

2.17 The partial ending of the franchise in **Scotland** in April 1994 led to about 5% of non-franchise customers switching supplier, mostly to the other Scottish PES, although a small number contracted with second tier suppliers south of the border.

COMPETITION IN GAS SUPPLY MARKETS

2.18 Competition has been, and will continue to be, introduced into the supply of gas (see Chapter 10). Prior to 1990, the gas market was dominated by British Gas (BG). By December 1994, the Office of Gas Supply (OFGAS) reported that there were 42 independent gas marketing companies (third party shippers) recording nominated gas supplies in the UK. OFGAS estimates suggest that, by the end of 1994, third party shippers had about half of the "contestable" gas market (i.e. supplies of 2,500 therms and over per annum,

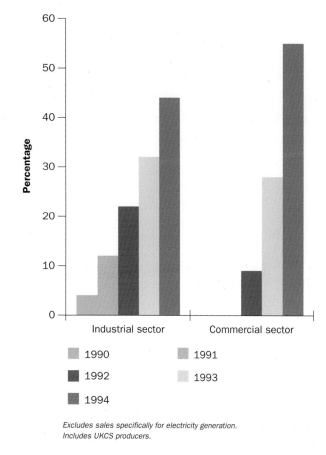

Chart 2.5

Independent suppliers' share of the industrial and commercial gas markets, 1990(Q4) to 1994(Q4)

Excludes sales specifically for electricity generation. Includes UKCS producers.

Source: Department of Trade and Industry

excluding power stations, chemical feedstocks, and natural gas vehicles) and they had thus exceeded the target set by the competition authorities of a 45% market share by 1995 (see chapter 10). This compares with less than 20% at the end of 1992, and only about 1% of a lower contestable total at the end of 1990.

2.19 Data on suppliers' shares are broken down into the categories of users – industrial, commercial, and power generation. Chart 2.5 shows the independent suppliers' share of the industrial and commercial sectors in the final quarter of each of the years 1990 to 1994. As a result of BG's agreements with the competition authorities, the independents' share has risen substantially over the last few years.

2.20 Competition is not evenly distributed. New entrants have about three quarters of the "firm contract" market (supplies which the seller is contracted not to interrupt), but there is currently very little alternative to BG in the "interruptible" market (sale of gas under arrangements which, if necessary, allow for the supply to be cut off for a number of days each year to assist in the balancing of total supply and demand). The independents dominate the supply of gas to gas-fired electricity generating plant: in the fourth quarter of 1994 they were responsible for about three-quarters of supplies.

Conclusion

2.21 There have been numerous new entrants to both the industrial gas and the industrial electricity markets. Many customers are now buying supplies from them. Even so, there remain areas where the previous incumbent firms remain dominant – one example is the supply of "interruptible" gas. There has been a significant amount of entry into electricity generation, but the market structure remains of concern to the regulator.

ELECTRICITY AND GAS PRICES

2.22 The aim of privatisation and liberalisation is to improve the service to both large and small customers. Inevitably, gains will occur at different times in different parts of the market. Customers will not always see immediate improvements. Indeed, in some cases, the unwinding of the various distortions introduced under nationalisation will have worsened the position of some groups (see paragraphs 2.27-2.29). But there is no doubt about the general direction of change.

2.23 This section discusses trends and changes in real prices. These are prices from which the effects of inflation, as measured by the Gross Domestic Product (GDP) deflator at market prices, have been removed. The main comparisons exclude the effect of Value Added Tax (VAT), which applied to domestic sector prices from 1 April 1994, since this was not the companies' responsibility. But some details are included of the movement of prices which include VAT.

Electricity prices

2.24 The scope for reductions in costs to feed through into reductions in prices depends in part on the importance of the different kinds of cost for supplies to the different sorts of customer. OFFER's estimates for typical annual bills for customers as a whole (both industrial and domestic) in 1993/94 in England and Wales show that generation costs account for over half of the price, distribution costs for about one-quarter, the fossil fuel levy for about 10%, and the remainder equally split between the costs of transmission and of supply. On average, the cost of primary fuels (coal, gas, and oil), which is included in generation costs, constitutes about 20% of the average bill.

2.25 Chart 2.6 shows the trends in average electricity prices to domestic and industrial customers, in index form, before

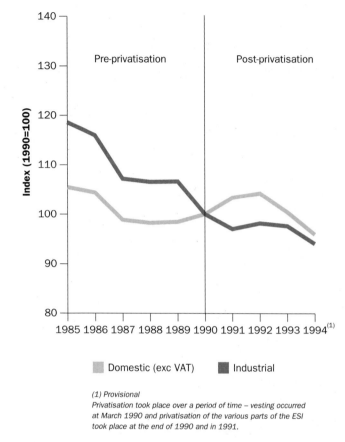

Chart 2.6

Trends in real industrial and domestic average electricity prices, UK, 1985 to 1994

(1) Provisional
Privatisation took place over a period of time – vesting occurred at March 1990 and privatisation of the various parts of the ESI took place at the end of 1990 and in 1991.

Source: Department of Trade and Industry

and after privatisation. Prices fell during the mid-1980s, partly due to the fall in fossil fuel prices, and then remained broadly constant over the three years leading up to privatisation.

2.26 Following privatisation, the average domestic price initially increased, up to 1992, but has since fallen to below its 1990 level – the average price is now (Q4 1994) 5% below that at privatisation (Q2 1990), though it is 2½% above when VAT is included. The average industrial price fell in the first year following privatisation. It then increased slightly, before falling in each of the last two years, with the fall in 1994 being much larger than that in 1993 – the average price is now (Q4 1994) 5½% below that at privatisation.

2.27 Prices to all groups of industrial consumers, from "small" users to "extra large" users, fell in 1994 (see Table 9.3). These figures are averages that conceal wide variations for individual companies. This has been particularly noticeable for large users over the last few years; some companies have incurred sizeable real increases in price, while others have benefited from large real decreases. Since 1991/92, some extra large customers have experienced price increases of over 30% in real terms. Some non-franchise customers have been paying prices based on those in the electricity Pool, which have fluctuated substantially (see Chart 9.1). However, most non-franchise customers buy electricity at a specified price – for example, on one-year or two-year contracts, or on terms which are related to Pool prices but with less variability.

2.28 An indication of the variation in prices for individual companies can be seen by comparing the lowest price paid by manufacturers in Great Britain (as evidenced by the lowest decile of the distribution of prices) and the highest price (the highest decile). In 1994, the highest price, on a quarterly basis, was 75% to 100% above the lower price, with the largest differences being in the winter period. The DTI's *Digest of UK Energy Statistics* and *Energy Trends* give more detailed figures of the distribution of prices paid.

2.29 Between 1989/90 and 1993/94, the overall level of electricity prices was largely pre-determined by the combination of the fuel supply and power sales contracts put in place at vesting and the initial "X" values in the various ("RPI±X") price controls. Very large users also benefited from the one-year transitional scheme (known as the Large Industrial Consumers Scheme, LICS) put in place at vesting, which ensured that their prices rose by no more than the rate of inflation in that year (1990/91). This helps to explain why prices to certain extra-large consumers fell in 1990 before rising in subsequent years. One reason for the increase in the price to the extra large customers was the ending of the favourable treatment they received before privatisation under the Qualifying Industrial Consumers Scheme (QUICS). LICS and QUICS did not apply in Scotland.

2.30 Over this initial period of privatisation, the electricity industry as a whole benefited both from a markedly faster increase in productivity than expected and from keener primary fuel prices. Wage and salary costs in the industry as a whole fell by around 10% in real terms over the period, with employment having fallen by about one-quarter. Coal prices to the generators fell by about one-third in real terms. Trends in the spot price of coal to power stations are shown in Chart 8.4.

2.31 From 1990 onwards, all electricity prices in England and Wales contained the Fossil Fuel Levy, which accounts for about 10% of the prices paid by all consumers. It was needed mainly to meet the higher than anticipated back-end costs of nuclear power revealed during the privatisation process.

2.32 Since privatisation, there has been a tightening in the price regulation: the transmission price cap, (X in the RPI-X formula), increased from 0 to 3 from April 1993 (i.e. price increases had to be held to 3 percentage points below inflation), and the supply price cap increased from 0 to 2 from April 1994. From April 1995, the price caps were tightened for the distribution businesses: distribution costs account for a much larger proportion of total costs than supply and transmission costs. The initial price caps allowed increases above the RPI for some RECs, the price formula ranging from RPI+0 to RPI+2½. In 1995/96, the RECs' price caps will be between RPI-11 and RPI-17, and will then continue for each of the next four years at RPI-2. These proposed price reductions are equivalent, in present value terms, to price controls with constant five-year values equivalent to between RPI-5½ and RPI-7½. The DGES said in March 1995 that he would implement these distribution price caps from 1 April 1995, but that he would consider whether there should be a further tightening of these price caps from April 1996 (see paragraph 9.28).

Gas prices

2.33 As in the case of electricity, it is useful to start with the breakdown of costs. Recent estimates by OFGAS and by BG show that the costs of gas supplies and of transportation/storage each accounted for about 40% of gas prices to domestic customers in 1994. The remainder was accounted for by supply (trading) costs.

2.34 Chart 2.7 shows the trends in average prices to domestic and industrial customers, in index form, before and after privatisation. It shows a fairly continuous fall in gas prices, a trend that started (broadly) in 1982 for industrial customers and in 1984 for domestic customers. The average price is now (Q4 1994) 39% below that at privatisation (Q4 1986) for industrial consumers, and (Q4 1994) 23½% lower for domestic consumers (17½% lower when VAT is included).

2.35 Much of UK gas is bought on individually negotiated field-by-field contracts, with the purchase or beach price applying for a full contract year (October-September). The initial price and the price adjustment formula relate the beach price to the price of competing fuels in the final markets, such as fuel oil and gas oil. Since 1985, the beach price has fallen, after a lag, in response to the fall in sterling fuel oil and gas oil prices (see Chart 10.2).

2.36 Since privatisation, as in the case of electricity, there has been a tightening in price regulation. Between 1986 and 1992, the price formula was RPI-2, and was set at RPI-5 for 1992-97.

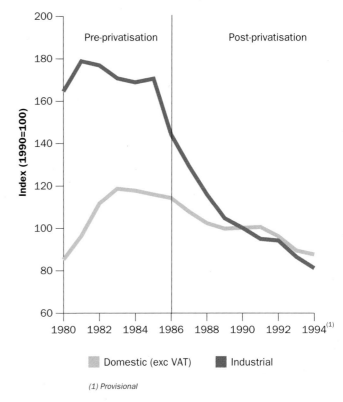

Chart 2.7

Trends in real industrial and domestic average gas prices, 1980 to 1994

(1) Provisional

Source: Department of Trade and Industry

COMPETITION IN ENERGY MARKETS

It was subsequently revised in April 1994 to RPI-4, following the 1993 MMC report (see paragraph 10.9), which recommended this change because the competitive threshold had been reduced from 25,000 therms to 2,500 therms in August 1992.

Conclusion

2.37 Prices have generally fallen in real terms since privatisation, substantially in the case of gas. In the electricity market, substantial reductions in prices are in prospect over the next few years under the new price control formulae which the companies have agreed with the DGES. It is as yet unclear whether the DGES will wish to tighten the price control formulae from April 1996.

MARKETS IN TRANSITION

2.38 Liberalisation and restructuring in the electricity and gas industries have already led to diversification into new businesses, the internationalisation of activity, and the growth of consultancy activities, joint ventures, direct equity, and asset investment. Some of the overseas activities of the gas and electricity industries are mentioned in chapter 7.

2.39 Within the UK, all the PESs are now involved in supplying gas to small/medium industrial and commercial customers, and many have pursued some degree of vertical integration by becoming involved in generation. The major generators have also integrated into fuel supply markets, especially gas: for example, National Power has stakes in an oil and gas development company, an oil consortium, and a North Sea gas field. NGC has moved into telecommunications and related businesses: a subsidiary provides telecommunications via fibre optic cables along NGC's transmission network. ScottishPower and Hydro-Electric are also investing in fibre optic networks.

2.40 There is already unrestricted competitive access to electricity generation. The pattern of entry will depend in part on the kind of contract cover which new entrants can obtain to offset their risks, and in part on the constraints presented by environmental safeguards. New entrants account for a large part of the recent additions to capacity, and many of these entrants have long-term contracts with the RECs (which sometimes part-own them). This could be taken to imply that entry depends on the existence of long-term contracts, but two recent CCGT projects have not been wholly backed by 15-year power purchase contracts with RECs.

2.41 Competition for new gas supplies from the UKCS will be determined largely by the way in which competition develops in final gas markets. Demand will be affected both by the obligations imposed on gas shippers under BG's Network Code (rules designed to ensure continuity of supply and access on non-discriminatory terms) and by the extent to which new market structures develop to support competition. A key development would be a "spot" market for gas, which would enable shippers to match supplies and demands far more efficiently than would be possible if they had always to rely only on their own contracted purchases.

2.42 The next stages in the development of competition in retail electricity and gas markets are set out in the "milestones" box. These lead towards 1998.

At that time, the Government plans to allow free access to all markets for the supply of gas and electricity to final consumers, but with regulatory and other structures embodying a number of important safeguards to ensure that competition will not jeopardise key social objectives.

2.43 Access to industrial and commercial customers is already largely competitive, but there are doubtless more changes to come in market shares and participation. The shape of competition in the domestic market will take a while to emerge post-1998. If the franchise sectors follow the precedent of the competitive electricity market, suppliers will compete for domestic customers outside their geographical areas (i.e. become second tier suppliers) when the franchise market ends in 1998. There is also the prospect of electricity and gas supply companies competing for the same business.

2.44 This is potentially a fundamental change. Domestic consumers are often fairly passive in their purchases of energy. Most people make some effort to keep their bills down, perhaps by investing in insulation or

PRIVATISATION AND LIBERALISATION OF THE ELECTRICITY AND GAS INDUSTRIES: MILESTONES

Electricity

- *Vesting*

  March 1990: CEGB split into National Power and PowerGen (fossil fuel generation), Nuclear Electric (nuclear generation) and the National Grid Company (transmission). 12 Regional Boards become separate companies (RECs).

  March 1990: South of Scotland Electricity Board and North of Scotland Hydro-Electric Board replaced by ScottishPower and Scottish Hydro-Electric (generation, transmission, supply, and distribution) and Scottish Nuclear (nuclear generation).

  March 1992: Northern Ireland Electricity replaced by 4 independent power stations (generation) and Northern Ireland Electricity plc (transmission and distribution).

- *Flotation*[1]

  December 1990: The 12 RECs.

  March 1991: 60% of National Power and PowerGen. The remaining 40% Government shareholding sold **March 1995**.

  June 1991: ScottishPower and Scottish-Hydro Electric.

  June 1993: Northern Ireland Electricity plc.

  (1) Nuclear Electric and Scottish Nuclear remain in the public sector. National Grid Company shares owned by RECs, although RECs have no policy control over NGC's transmission business. At flotation the Government retained a special share in all floated companies, giving it the power to be consulted in the event that a shareholder's holding reached more than 15%. Special shares in each of the RECs were redeemed on 31 March 1995.

- *Competition*

  Number of customers having freedom to choose suppliers gradually widened:

  March 1990: At vesting, freedom only for those with maximum demand over 1 MW.

  April 1994: Limit reduced to 100 kW.

  1 April 1998: "franchises" to end – limit to be abolished completely.

Gas

- *Vesting*

  August 1986: The assets of the British Gas Corporation vested in British Gas plc.

- *Flotation*

  December 1986: Government retains a special share in the company.

- *Competition*

  December 1986: At privatisation, British Gas retained monopoly in the tariff market. It had lost its legal monopoly in the non-tariff or contract market (customers with demand greater than 25,000 therms per year) in 1982, but had effectively retained this monopoly until privatisation.

  1988 MMC Report on contract gas market.

  March 1992: British Gas undertook to the Director General of Fair Trading to create the conditions by which competing suppliers should be able to supply at least 60% of the market above 25,000 therms (subsequently redefined to 45% of the market above 2,500 therms). By December 1994 competitors had about half of this (redefined) market.

  August 1992: Market for demand between 2,500 and 25,000 therms per year opened up to competition.

  July 1993: MMC Report on competition in gas supply.

  1 April 1996: 1 April 1997; 1 April 1998:

  Government plans a phased introduction of competition to the domestic market.

  1 April 1996: First tranche of liberalisation: approximately 500,000 customers in a single region.

  1 April 1997: Second tranche: competition extended to cover 2 million customers.

  1 April 1998: Competition extended to all UK customers.

other energy-saving devices, and they may assess the merits of different fuels when they buy significant new items of equipment – cookers, central heating boilers, fires etc. But they tend not to review their options regularly. There is now the possibility of significant change, with energy companies making a much more coherent effort to tell their customers about the merits of particular fuels in particular uses, and even to present them with a range of options for saving energy. Monopoly suppliers would seem to have little interest in getting their customers to review their choices once they are committed to a particular fuel. Competition, on the other hand, could encourage suppliers to meet the needs of their customers by offering a full range of energy services, including energy efficiency advice, for fear that failure to meet these needs will induce customers to change suppliers.

2.45 One constraint on the development of competition may be the extent to which new contractual arrangements can be developed. In the 100 kW-1MW electricity market, there currently has to be a supply agreement between a customer and a supplier, and a connection agreement between the customer and the local REC. Domestic customers, too, are likely to be offered new sorts of contract. Experience with other industries suggests that competition is likely to increase the range of different tariff packages on offer. If so, some customers will find packages which reduce their bills, compared to the standard tariff. It is even possible that some larger users may find it worthwhile to buy through an intermediary who can shop around on their behalf and offer the domestic equivalent of an industrial energy management service.

2.46 A key element in the process of change is the development of metering technology. There is already a wide range of meters available, and few technological limits on the possible uses of information technology to increase the range of metering services. In principle, "smart" meters could seek out the cheapest supplier at any particular time of day, and control a household's use of energy far more precisely according to the cost of units – for example, by switching on washing machines, dishwashers, or tumble driers during the night.

2.47 The obvious constraint is the cost of buying and fitting a meter. For industrial electricity supply, the annual charges for second tier metering have fallen from £5,000 per meter in 1990/91 to £1,140 in 1993/94, and have been set at £200 for each 100 kW customer from 1994/95. A recent OFFER paper (see paragraph 2.51) considered alternative possibilities for upgrading meters and simpler alternative arrangements for access by second tier suppliers.

2.48 Although the cost of a meter has fallen markedly, it is still high in relation to the average domestic bill of £300-500 a year. But some people believe that meter manufacturers will eventually be able to deliver a domestic meter for around £50. If so, this might encourage more customers to consider switching suppliers.

2.49 The degree to which householders are likely to involve themselves actively in energy "trading", via smart meters or supplier switching, will depend, of course, on the potential for savings. This in turn depends on the effectiveness of the new entrants. Given that the costs of supply are a relatively small proportion of total costs, the scope for using competitive pressures to reduce costs is, at first sight, limited. Moreover, the existing price controls already act as a proxy for some of the pressures of a competitive market. But it is never easy to see in advance how competitive markets will operate, and quite where efficiency can be improved. In both the gas and electricity markets, one source of costs reductions could be improved system efficiency, particularly if prices more closely reflect the costs of supplying particular types of users. Another possibility is that new suppliers will strike different contracts with suppliers of primary energy. Certainly in the case of the firm gas market, price reductions have already been sizeable, with competitors offering prices 10% or more below the BG schedule price. Some leading independent companies believe that these sorts of price reductions will occur in the domestic sector when it is subject to competition.

2.50 The process of change inevitably calls into question the ways in which the industries and their regulators will continue to provide for the continuing needs of particularly vulnerable groups of customers, such as the disabled. These issues must be addressed, but they are not an insuperable problem, and arrangements can be put in place to protect these groups. The details of the Gas Bill show how this will be done in the case of gas (see chapter 10).

2.51 The debate about the appropriate arrangements for electricity has started with the publication of the DGES's consultative document on 1998 (see Appendix 5). This sets out his initial observations on the trading and licensing arrangements which might be possible after the electricity supply market becomes fully competitive. He has set up a 1998 Co-ordination Group to advise him on the development of appropriate arrangements. The first meeting of the Group took place in March 1995. He has also written to the Electricity Pool, asking it to report to him by the end of April 1995 on the options for trading arrangements in 1998.

2.52 As suggested above, there is the prospect that competition will encourage companies to offer a full range of energy services, including energy efficiency advice. But whatever the activities of private companies, the Government will continue to promote the efficient use of energy both during and after the introduction of competition into the gas and electricity industries. It will review how far competition is of itself enough to realise efficiency, and consider any further steps needed to promote energy efficiency. It will also consider the extent to which assistance should be provided to those who cannot afford to finance energy efficiency measures.

Conclusion

2.53 The liberalisation of the domestic gas and electricity market, and suppliers' commitment to the principles of the Citizen's Charter, offer the potential for significant improvements in the services offered to customers. The continuing needs of special groups of customers can be addressed within the accompanying regulatory structures.

CHAPTER 3

ENERGY COSTS, ENERGY EFFICIENCY, AND INDUSTRIAL COMPETITIVENESS

3 Energy Costs, Energy Efficiency, and Industrial Competitiveness

3.1 Energy prices in the UK have been falling in real terms (see Chapter 2). A reduction in costs is one factor in improving the competitiveness of UK firms (see box). A reduction in energy costs will be achieved by a combination of a fall in energy prices, relative to those in other countries, and an increase in energy efficiency.

3.2 This chapter looks at how energy feeds into the competitiveness of UK firms. It is in three parts. It begins by examining the

> The 1994 *Competitiveness White Paper* defined competitiveness for a firm as "the ability to produce the right goods and services of the right quality, at the right price, at the right time. It means meeting customers' needs more efficiently than other firms". For a nation, the OECD defines competitiveness as "the degree to which it can, under free and fair conditions, produce goods and services which meet the test of international markets, while simultaneously maintaining and expanding the real incomes of its people over the long term".

Chart 3.1

Energy costs as a proportion of production costs, 1989

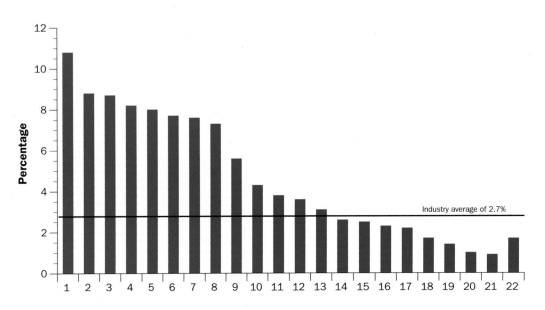

Key: 1 = Water supply; 2 = Manufacture of paper & board; 3 = Bricks, ceramics etc; 4 = Ferrous foundries; 5 = Basic chemicals; 6 = Cement, lime & plaster; 7 = Extraction of minerals; 8 = Glass & glassware; 9 = Iron & Steel; 10 = Non-ferrous foundries; 11 = Non-ferrous metals; 12 = Other chemicals; 13 = Rubber & Plastics; 14 = Manufacture of textiles; 15 = Food, drink & tobacco; 16 = Manufacture of metal goods; 17 = Paper conversion; 18 = Manufacture of vehicles; 19 = Engineering; 20 = Printing & publishing; 21 = Clothing & footwear; 22 = Others

Source: Energy Paper 64: Industrial Energy Markets

ENERGY COSTS, ENERGY EFFICIENCY AND INDUSTRIAL COMPETITIVENESS

importance of energy to the UK economy and to industry; it then compares energy prices in the UK with those in other countries; and finally it considers the importance of energy efficiency.

ENERGY AS A COST

3.3 Purchases of energy account for about 5% of the UK's GDP, but their importance to industry varies. A recent DTI publication (Energy Paper 64 (EP 64), *Industrial Energy Markets*) shows energy costs as a proportion of production costs for the main manufacturing industries (see Chart 3.1 and Table A17), as well as for more than 200 sub-divisions of manufacturing industry.

3.4 Chart 3.1 shows that for all manufacturing industries, energy costs accounted for an average of 2.7% of production costs, ranging from a high of about 11% for water supply to a low (for an individual grouping) of about 1% for clothing and footwear. Nine of the 22 sectors had energy costs of above 5% of production costs.

3.5 Between 1985 and 1989, energy costs fell as a proportion of production costs for manufacturing industry, both as a whole and in every sector. This was the result of a combination of structural change, falling oil prices and increased energy efficiency. The overall fall was about 1%, but the largest reduction for a sector was around 6% (glass and glassware).

UK ENERGY PRICES COMPARED WITH THOSE IN OTHER COUNTRIES

3.6 In this section, the prices of electricity and gas to industry in the UK are compared with those in six of the UK's major trading partners (Table A24 shows a comparison with all other countries in the EU and with some other countries in the OECD). Movements in exchange rates affect UK firms' prices relative to those of other countries. Between 1991 (the last full year before the UK left the ERM) and 1994, sterling fell by about 34% against the yen, by about 15% against the franc and the mark, and by 13% against the dollar, but rose about 12% against the lira and the peseta. Reductions in sterling reduce the price of UK goods and services relative to those of other countries, while increases in sterling increase the relative prices of UK goods and services.

3.7 Charts 3.2 and 3.3 overleaf use 1993 exchange rates and compare average industrial electricity and gas prices in 1993 for all the chosen countries except for Spain (where the latest available gas data are for 1992).

3.8 The prices in Charts 3.2 and 3.3 include non-refundable taxes, i.e., they exclude VAT. For electricity, UK prices lie towards the bottom of the range, with only France and the USA having lower industrial electricity prices. For gas, UK prices currently lie towards the bottom of the price range (it is sixth out of 7). Including taxes makes no difference to the relative position of the UK.

Chart 3.2
Average industrial electricity prices, 1993

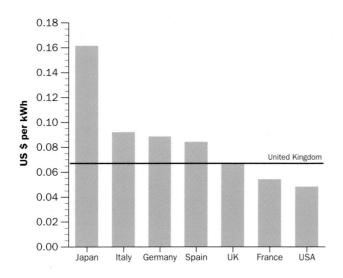

Excludes refundable taxes, e.g., VAT.

Source: IEA Statistics – Energy prices and taxes, 1994 (Q3)

Chart 3.3
Average industrial gas prices, 1993

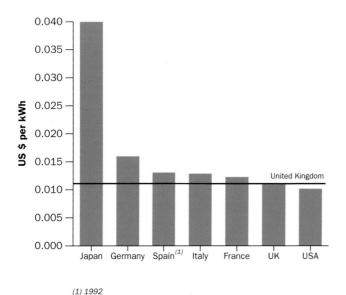

(1) 1992
Excludes refundable taxes, e.g., VAT.

Source: IEA Statistics – Energy prices and taxes, 1994 (Q3)

ENERGY COSTS, ENERGY EFFICIENCY AND INDUSTRIAL COMPETITIVENESS

3.9 Over a five year period, the relative movements in both exchange rates and electricity and gas prices (as expressed in US dollars) have improved the relative prices of UK firms compared with firms in all these countries except Spain. Where prices in other countries were higher than those in the UK in 1988, these differences have increased; where prices were lower, these differences have been reduced or reversed.

3.10 There are other ways to measure relative prices. Information about prices categorised by size of customer is published by the European Statistical Office (Eurostat) in its *Rapid Reports on Energy and Industry*. In Charts 3.4 and 3.5, average electricity and gas prices are shown for seven and five sizes of consumer respectively for five European countries. The charts refer to July 1994, the latest available data.

Chart 3.4

Average industrial electricity prices by size of customer, July 1994

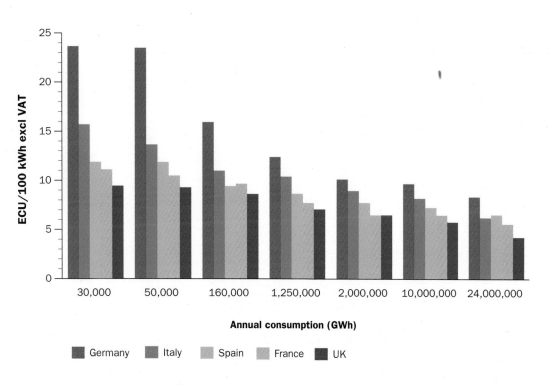

The following data are used: Germany – Frankfurt, Italy – Milan, Spain – Madrid, France – Lille, UK – Birmingham.

Source: Eurostat: electricity prices, various editions.

3.11 For electricity, the UK has the lowest prices for all but one of the size bands shown. During the last few years, the relative movements in exchange rates and electricity prices have improved the relative prices to UK firms, compared with firms in Germany and France, in all of the above categories. The comparison with prices in Spain and Italy is less straightforward: in some categories the relative position of UK firms has worsened. Chart 3.4 includes seven different size categories, but it does not cover extra-large users: the largest category is at the lower end of the range for "moderately large" consumers as defined in the DTI's own series (see DTI's *Energy Trends* for details). It is very difficult to compare data for extra-large users on a like-for-like basis, due to the existence of special terms within contracts. However, existing data suggest that prices in the UK for these extra-large customers are higher than those in most European countries, with the exception of Germany and, to a lesser extent, Spain. The lack of coverage of the larger users explains why France has higher average prices than the UK in virtually all of the consumption categories shown in Chart 3.4 and yet a lower overall average price in Chart 3.2: France has lower prices for larger users.

Chart 3.5

Average industrial gas prices by size of customer, July 1994

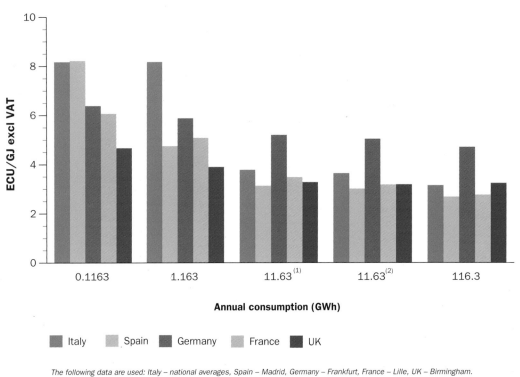

The following data are used: Italy – national averages, Spain – Madrid, Germany – Frankfurt, France – Lille, UK – Birmingham.
(1) Utilization 200 days and 1,600 hours.
(2) Utilization 250 days and 4,000 hours.

Source: Eurostat, Gas prices.

ENERGY COSTS, ENERGY EFFICIENCY AND INDUSTRIAL COMPETITIVENESS

3.12 In the case of gas, UK prices are the lowest or close to the lowest for most of the categories shown. In the largest category, the average UK price is towards the upper end of the range, but it is significantly below that in Germany. Given movements in exchange rates and gas prices since 1988, UK firms have experienced an improvement in their relative prices in virtually every consumption category shown above.

ENERGY EFFICIENCY

3.13 The use of energy is not an end in itself: its purpose is to provide a range of services, such as heat, light, transport, and power for machines and appliances. By increasing the efficiency with which energy is used, firms can reduce their unit costs and thus improve their competitiveness. They can also reduce their exposure to fluctuating market prices for energy.

Trends in energy efficiency

3.14 It is useful to look first at who uses energy (see Appendix 1 for further details of final energy consumption). In 1994, the transport sector was the largest user of final energy (33%), followed by the domestic sector (29%) and the industry sector (24%), with the services sector accounting for the remainder (14%). In recent years, there has been very little change either in total final energy consumption or in consumption in each of the sectors. Taking a longer view, however, there have been substantial changes. Since 1980, total final consumption has increased by about 8%. Within the period, the use of energy in the transport sector has increased by 42%, with most of the change coming from road transport and most of the remainder from air transport. Over the same period, industry's energy use has fallen by about a quarter, due in part to the declining importance of this sector to the UK economy: the decline was particularly severe in the energy-intensive industries. The growth in energy consumption in the domestic and services sectors was slightly higher than that of the overall total.

3.15 The importance of individual fuels differs amongst the sectors. Petroleum accounts for virtually all of the fuel used in the transport sector. In the domestic sector, gas is the main fuel (64%), due to its popularity for space and water heating. In these two sectors, there has for some time been little change in the major fuel used. Gas (32%), electricity (23%), and petroleum (23%) are the major fuels in the industry sector. Over time, the share of gas has remained the same, but electricity's share has increased, largely at the expense of petroleum.

3.16 Some indication of changes in energy efficiency can be gained from movements in the "energy ratio" (see box overleaf). A falling ratio, as shown for most of the period since 1970 in Chart 3.6 (overleaf), is consistent with an increase in energy efficiency. But there are other reasons for changes in the energy ratio. One such reason is the movement away from energy-intensive activities, such as steel making, towards low energy activities, such as services. The energy ratio is defined in terms of primary energy rather than final energy, and so it includes improvements in the efficiency of energy production – most notably in the efficiency of power generation – as well as greater efficiency in its use.

The **ENERGY RATIO** shows the relationship between primary energy demand and aggregate economic activity. It is calculated as follows:

Energy Ratio = $\dfrac{\text{total inland consumption of primary energy (temperature corrected)}}{\text{gross domestic product at 1990 factor costs}}$

Chart 3.6 shows the trends in the energy ratio and its components between 1970 and 1994.

Chart 3.6
Trends in the energy ratio and its components, 1970 to 1994

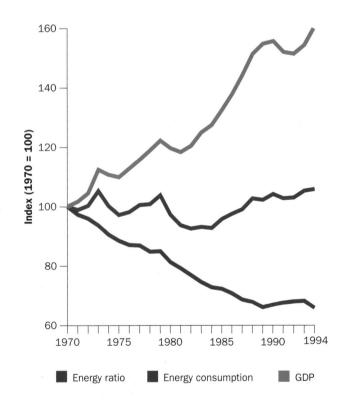

Source: Department of Trade and Industry

The chart shows that, apart from a small increase in 1979, the energy ratio fell in every year between 1970 and 1989. Between 1989 and 1993, at the time of the recession, it increased gradually before declining in 1994. The rise in the energy ratio between 1989 and 1993 was partly due to the effects of the recession. A good deal of energy consumption, such as that required for heating and lighting buildings, does not change with changes in economic activity. In previous recessions, the energy ratio continued to decline, as the effects of those recessions were felt most severely in energy-intensive industries.

3.17 Another method of estimating the improvement in energy efficiency in industry is to attempt to disaggregate the reasons behind observed changes in final industrial energy consumption. Although industry accounts for only about one-quarter of final energy consumption, changes in industrial use have been responsible for most of the variation in overall energy use over the past two decades. The results of such an analysis were published in EP 64 (see paragraph 3.3), which analysed the change in industrial energy consumption between 1973 and 1993 in terms of three effects: an energy efficiency effect, an output effect, and a structural change effect.

3.18 Because of the limitations of the data sources used, the results of this exercise are less reliable in the years before 1979 and, to some extent, since 1989 than in the decade from 1979 to 1989 (nearly three-quarters of the fall in energy use between 1973 and 1993 occurred between 1979 and 1989). Care is needed in interpreting the energy efficiency effect. It includes the effect of scrapping old equipment, which is likely to be less energy-efficient, and its replacement with new and more efficient plant, as well as the effect of introducing specific energy efficiency measures and techniques. It may also include some small effects from changes in the pattern of production at a more detailed level than the aggregate industrial statistics permit. Nevertheless, some broad conclusions can still be drawn from this work. Between 1973 and 1993, improved energy efficiency seems to have accounted for a large part of the fall in energy consumption by manufacturing industry. The gains in energy efficiency varied considerably over this period, and were greatest between 1979 and 1984, when energy prices were high. Further gains were made between 1984 and 1989, even though energy prices fell sharply in 1986. The data suggest that, contrary to the experience in earlier periods, the fall in consumption between 1989 and 1993 was primarily due to output and structural change effects, and not to energy efficiency. This suggests, unsurprisingly, that the incentive to improve energy efficiency is greater when energy prices are high than when they are low.

Energy efficiency: the Government's approach

3.19 Energy prices can therefore provide an incentive to be more efficient in the use of energy. The gradual removal of cross-subsidies, and the more accurate reflection of costs, should promote energy efficiency, but there are factors which may prevent firms and households from making sensible investments. Thus, lack of information is generally accepted as a barrier to energy efficiency. Equally, the market for energy efficiency services is still undeveloped – most obviously in domestic markets, but also in some industrial markets. Other factors are also important, such as the separation of those who receive the benefits of energy efficiency from those who incur the costs (e.g. the separation between office leaseholders and the developers or owners). Lack of investment can be a problem. There is also a disparity in the ability to borrow between the corporate sector, which has traditionally invested in the supply of energy, and the domestic sector.

3.20 The Government's broad approach is to support the market by promoting information and awareness. Information is a necessary, if not a sufficient, condition for markets to work properly. There are relatively few cases where individual decision-makers are likely to be in a position to establish all the necessary information themselves.

3.21 Information is supplemented selectively in three important respects: regulation, particularly where key decisions are to be made; financial support, particularly where individuals are concerned; and direct influence on policy decisions within the public sector itself. The rest of this chapter reviews policy, first in the industrial and commercial sectors, where competitiveness is a major issue; and then in the public and domestic sectors, where similar policies have been developed.

Energy efficiency and competitiveness

3.22 The Government's *Making a Corporate Commitment Campaign* encourages top management to commit themselves to improving energy efficiency. It asks them to formulate, adopt, and publish a corporate policy on energy efficiency; to increase awareness of the benefits of energy efficiency amongst their employees; and to set targets for improving their performance. So far, more than 1,800 organisations have declared their commitment. Feedback indicates that the campaign is helping to raise the profile of energy management and to provide a focus for action. Several organisations, stimulated by the campaign, have made very large savings (see box). Experience suggests that an organisation that takes energy efficiency seriously also takes other aspects of its business –

Stimulated by the Making a Corporate Commitment Campaign:

GPT Ltd, a manufacturer of telecom equipment, has:
- made fuel savings of £1.4 million
- reduced CO_2 emissions by 30,000 tonnes
- made average water savings of 67% at each of its sites.

Northumbrian Water has:
- made energy savings of £850,000 since 1990/91
- reduced net annual energy costs by 17%.

British Sugar has:
- saved over £10 million in ten years.

---

environmental management, product quality, and health and safety – seriously. This suggests that lenders will be well advised to look at how a company addresses these issues when taking key decisions.

ENERGY COSTS, ENERGY EFFICIENCY AND INDUSTRIAL COMPETITIVENESS

3.23 A recent study by Touche Ross for the Department of the Environment (DoE) confirmed that high energy-intensive companies treated energy efficiency as a strategic issue, subjecting proposals to the same investment criteria as the rest of their core business. But less energy-intensive companies tended to treat energy efficiency as a discretionary item, requiring a high return or a quick payback. Such companies had not seen energy as important, and some had not established whether energy efficiency opportunities existed. Companies with a "best in class" philosophy, however, tended to give higher priority to energy efficiency investment. The Corporate Commitment Campaign approach, aimed at the boardroom and at creating awareness at the top of the office, seeks to extend this approach more widely.

3.24 The Government provides a range of information and technical advice to help organisations of all sizes improve their energy efficiency and hence their competitiveness. The DoE's main technical information and research and development programme is the *Best Practice Programme*. This provides energy users and managers and building professionals and managers with authoritative and tested information on how to become more energy efficient.

3.25 The Best Practice Programme improves market intelligence through its energy consumption guides for industrial sectors, processes, and building types; encourages replication by promoting good practice; stimulates confidence in new energy efficiency measures by providing information on their performance; and secures financial support for research into new energy efficiency measures. On a conservative estimate, savings of £300 million a year had been stimulated by the end of 1994. The programme has an overall target of energy savings worth £800 million a year (at 1990 prices) by the year 2000.

3.26 There are many individual success stories – for example, the Rover group saved more than £1 million in 6 months by raising the awareness of employees about the need for energy efficiency. Several projects have reported on the successful installation of new industrial combined heat and power designs (see box).

> A combined heat and power plant installed by **English China Clays** with the help of the Best Practice Programme has:
>
> - given net cost savings of over 10% of the site's energy bill
> - used exhaust gases from the combined heat and power process to dry materials
> - reduced annual emissions of carbon dioxide by 44,000 tonnes
> - been so successful that the company is to install a similar scheme at another site.

3.27 These information programmes are supported by direct assistance for small firms, primarily designed to ensure they have bespoke information. The Energy Management Assistance Scheme has contributed to consultancy costs for small firms intending to improve their energy efficiency. This scheme will be phased out in its present form, and replaced by a grant to help small firms improve their overall environmental performance. This new grant, which will be known as the Small Company Environmental and Energy Management Assistance Scheme (SCEEMAS), will help manufacturing firms with fewer than 250 employees to address the environmental issues facing them and to participate in the EC Eco-Management and Audit Scheme. The grant will continue to facilitate energy efficiency as part of a company's overall environmental performance. It is expected to come into operation during May.

3.28 Through its Energy Design Advice Scheme, the Government provides expert consultancy advice to all those involved in the energy-conscious design of buildings, including architects, developers, surveyors, and local authorities. The scheme covers both new build and refurbishment projects, and provides a free consultation to identify possible design options which could improve the energy and environmental performance of the building. This consultation lasts up to one day, and is paid for by the Scheme. If the first consultation identifies significant potential savings which require more detailed examination, further consultancy may be offered. Up to 50% of these further costs may be reimbursed by the Scheme.

3.29 Participants have an opportunity to reduce their own and the nation's energy bills. In a typical office facility of 1,000m^2, where energy-conscious design and good practice have been implemented, annual savings of £5,000 can be realised. Experience of the Scheme to date suggests that, across the commercial and industrial sectors, savings of up to £50 million a year are possible. If achieved, this would cut emissions of greenhouse gases by nearly one million tonnes.

Energy efficiency: public and domestic sectors

3.30 The public and domestic sectors benefit both directly and indirectly from the Best Practice Programme, particularly the component which concentrates on energy efficiency in buildings. The public sector has also been stimulated by a series of initiatives, including:

- the campaign on the Government's own estate (which has also been picked up by other public sector organisations) to reduce energy use by 15% over a five year period;

- the Government's requirement that local authorities' housing investment programmes should include features designed to improve energy efficiency; and

- the production of comprehensive guidance on energy efficiency in council housing.

ENERGY COSTS, ENERGY EFFICIENCY AND INDUSTRIAL COMPETITIVENESS

A Private Member's Bill currently before Parliament (the Home Energy Conservation Bill) will, if passed, help to focus local authority activity by requiring them to produce reports identifying measures which would significantly improve the energy efficiency of residential accommodation in their area.

3.31 Domestic energy efficiency depends crucially on a large number of individuals and households taking a series of decisions on which they cannot be expected to be experts. The Government has run a series of campaigns to promote awareness of both the direct financial and the wider environmental benefits of energy efficiency. The *Wasting Energy Costs the Earth* campaign was launched in Autumn 1994, following on from the earlier campaign *Helping the Earth Begins at Home*. The new campaign:

- links energy saving and direct financial savings for the individual household;

- uses a family of dinosaurs (Ron, Brenda and little Bill) linked to a travelling roadshow aimed particularly at the young; and

- links the national campaign to individual product campaigns making use of the same dinosaurs and logo.

Both the electricity and gas regulators (OFFER and OFGAS) have recognised the importance of energy efficiency advice, and both require utility companies to provide it.

3.32 Advice is backed up by selective regulation. Building regulations have been progressively upgraded. The regulations made in July 1994 increased the energy performance of space and water heating in new dwellings by around 30% compared with previous provisions. They also required (for the first time) each new dwelling to receive an energy rating based on a standard assessment procedure (SAP). The SAP ranks each dwelling for its efficiency on a scale of 0-100. SAP rating for new dwellings will become mandatory in July 1995. The Government is in discussion with mortgage lenders about the prospects for promoting similar energy rating as part of house valuation for existing dwellings.

3.33 New EC regulations came into effect on 1 January 1995, requiring the labelling of fridges and freezers. This is expected to be the first of a series of such regulations which will ensure that consumers can make better-informed choices.

3.34 The main direct financial support to the domestic sector is provided by the Government through the *Home Energy Efficiency Scheme* (HEES). This provides grants to low-income households and to elderly and disabled people for basic energy efficiency measures – loft, tank, and pipe insulation, draughtproofing, and general energy advice. The scheme has proved popular, and during 1994/95 a significant waiting list developed, despite the doubling of the programme from £35 million to £70 million a year.

During the year, additional resources were made available, so that spending for 1994/95 will be some £77 million, and from 1995/96 onwards it will be over £100 million a year. A million homes have been treated so far under the Scheme, and the current level of funding will permit around 600,000 more to be treated each year. The DoE also part-funds a programme of action by the registered charities Neighbourhood Energy Action and Energy Action Scotland.

3.35 As has been suggested in chapter 2, the creation of more competitive structures in energy markets may well encourage the growth of private markets in energy efficiency services. With previous monopolistic public-sector suppliers, it is by no means clear that investments on the supply and demand sides of energy were on an even footing. Energy efficiency is likely to be promoted if there is equality. The process of adjustment to more competitive markets will take time, and will not stop once competition has been formally introduced. The development of better-informed markets and of particular services and products are therefore both likely to benefit from pump-priming. This is a key role of the Energy Saving Trust (see box opposite), which was set up jointly between the gas and electricity companies and the Government.

3.36 The Government is keen to encourage the establishment of energy service companies, within a sympathetic regulatory regime. Their aim would be to provide demand side management services for profit, thus serving both consumers and their own commercial interests.

THE ENERGY SAVING TRUST

The Trust is an independent non profit-making organisation set up by the Government, British Gas, and the Public Electricity Supply companies. It aims to develop and manage new programmes to promote the efficient use of energy by domestic and small business consumers. Its strategic plan, published in April 1994, anticipated major grant schemes, with funding raised mainly from levies on gas and electricity consumers.

OFFER has set electricity companies Standards of Performance on energy efficiency, which came into effect on 1 April 1994. It has also allowed companies to raise £1 per franchise customer each year for 4 years to finance energy efficiency projects. The Standards prescribe the level of energy savings which companies should achieve on behalf of customers, and lay down the criteria for projects. The Trust advised OFFER on the targets, and is responsible for endorsing projects for OFFER's approval and for evaluation and monitoring of the projects.

A pilot study (partly funded by Government) of some 30 Local Energy Advice Centres (LEACs) for the domestic and small business sectors has been very successful. In their first year, the LEACs have advised 26,500 clients, who have subsequently installed over 83,000 energy-saving measures. These have led to savings of around £1 million on fuel bills.

Two pilot schemes have been approved by OFGAS for funding by gas consumers:

- nearly 10,000 applications were made under a one-year condensing boiler pilot scheme, which offered £200 rebates;

- a two-year residential Combined Heat and Power scheme has been heavily oversubscribed. Grants of £1.3 million have been made so far.

Proposals for five new energy saving schemes developed by the Trust were submitted by British Gas in July 1994 to OFGAS for approval to pass the costs through to gas consumers. In February 1995, three of these were turned down, and further information sought on the remainder – a loan finance scheme for owner-occupiers, and a pensioner equity release project.

In the light of this, the Trust is now reviewing its plans and developing new and innovative ideas. The Government has taken powers to contribute to the running costs of the Trust, where appropriate, to enable it to carry forward these ideas.

CHAPTER
4

FUTURE UNCERTAINTIES

4 FUTURE UNCERTAINTIES

4.1 Chapter 2 flagged up some of the changes which are possible in the UK energy industries. These will depend on developments both on the supply side (largely the result of the next steps in the liberalisation of the energy industries) and on the demand side (as domestic and industrial consumers respond to wider changes in the economy and in society). But what happens in the UK will be strongly influenced by what happens internationally. We can expect, for example, the increasing internationalisation of trade and the further development of world capital markets. In the energy industries there are two over-riding international influences – the future course of oil prices and future international environmental agreements.

4.2 This chapter looks at some different visions of the future, and pays particular attention to the possible course of oil prices. It starts by taking an international view of the possibilities, and then narrows down to possibilities in the UK, drawing on new work within the DTI on UK energy projections. The possible effect on the UK of future agreements to limit emissions of acidic and greenhouse gases is covered in Chapter 5.

GLOBAL ENERGY FUTURES

4.3 A theme of this chapter is the uncertainty surrounding any attempt to think about the future. But there nonetheless remains a need to look ahead. There have been a number of attempts to draw out the possible pattern of development in world energy markets, notably the reports by the World Energy Council (WEC) and the International Energy Agency (IEA) (see publications list at Appendix 5). The conclusions of the IEA study are outlined in the box. The focus of both studies was on the pattern of world supply and demand. Two interests predominate: the extent to which it is possible to envisage imbalances between energy demand and energy supply, and the future course of world greenhouse gas (and particularly CO_2) emissions. Both studies make it clear that the need to respond to the different rates of economic growth in the developed and the developing world is at the heart of the debate about the appropriate response to the potential threat of climate change.

Energy prices

4.4 World energy prices have in recent times been driven by oil prices. These will always be difficult to predict, since they depend on a mix of political and economic factors. Present patterns of oil production do not reflect the relative cost advantages of different regions, and it is not always the cheapest oil which is exploited first. Nevertheless, spare capacity in the OPEC countries, coupled with the growing integration of world oil markets, has reduced the ability of any particular group of producers to manipulate market prices. Prices are therefore more likely than before to be established in relation to the underlying levels of demand and the marginal costs of supply. Even so, political

THE PRINCIPAL CONCLUSIONS OF THE INTERNATIONAL ENERGY AGENCY'S WORLD ENERGY OUTLOOK 1994 WERE THAT:

- world demand for primary energy will continue to grow, by about 50% from 1991 to 2010, driven by world GDP growth (up 70% over the same period);

- fossil fuels will continue to be the dominant energy sources;

- primary energy demand growth in OECD countries is projected to be 1.3% a year (compared to GDP growth of 2.3% a year), and a major component of growth will be oil for transport use;

- demand in the Rest of the World (ROW) countries (ie Middle East, Africa, Latin America, East Asia, China) will more than double by 2010, driven by rapid population growth, urbanisation, increasing industrialisation, increased economic and transport activity, and the diminishing availability of non-commercial sources of energy such as firewood;

- consumption of oil, gas, and coal will all rise; electricity will be the most rapidly-growing form of final energy, keeping pace with or exceeding GDP growth; nuclear electricity generation will not increase significantly;

- world carbon emissions will rise by 50% by 2010. Within this overall total, emissions from ROW nations will more than double, while OECD emissions will grow more slowly. This picture does not change significantly under alternative low economic growth assumptions;

- consequently, any international initiative to reduce CO_2 that is not adopted by non-OECD regions would be unlikely to have any major effect on global emissions, and an even smaller effect on atmospheric concentrations.

and other shocks are likely to produce periods of short-term price volatility. Since the end of the Cold War and the re-ordering of political power structures, there has been only one crisis directly affecting world energy supply – the Iraqi invasion of Kuwait. But the potential remains for further problems associated with political change elsewhere in a volatile world.

4.5 At present over 60% of supplies come from non-OPEC producers, even though these countries have only 20% of known reserves. Experience suggests that reserves estimates have little absolute meaning. A country which is producing at an annual rate equivalent to one-tenth of its reserves (as the US has always done) will not run out of oil in 10 years, since new reserves will be found. Indeed, the world as a whole has been finding new reserves faster than it has been consuming known existing reserves.

4.6 Over the next 10-15 years, the part which individual producers will play in meeting the increase in world oil demand will depend upon the scale of their reserves and the cost of increasing their production (relative to the oil price). In the past, many people have envisaged falling non-OPEC production. It now seems possible that total production from both OPEC and non-OPEC countries will rise, though within these totals not all countries will see an increase (for example, European production may be peaking). But whatever the case, it seems inevitable, given the distribution of reserves, that OPEC production will increase much more than non-OPEC production, such that OPEC will increase its share of the market.

4.7 Recent low oil prices have been beneficial for the world economy, and have helped to raise economic growth. But they have reduced the income of the producer nations. Over the past decade, OPEC producers have been meeting new demand by using spare production capacity. The increased demand from both OPEC and non-OPEC countries will entail new investment by both the oil companies and the producer nations. This implies that there will be an increasing call on the resources of countries which in many cases are already facing financial pressures. There must, therefore, be some risk that there will be insufficient investment to meet the growth in demand. If so, this would put upward pressure on oil prices.

4.8 Proven world oil reserves stand at about 50 years' production at current rates. The size of the economically-recoverable reserve is linked to prices. In the face of lower crude oil prices, the oil industry has achieved great success in reducing its costs both of exploration and development.

The success of the UK's Cost Reduction in the New Era (CRINE) initiative was noted in the 1994 Energy Report (see also paragraph 7.12 of this Report). Nevertheless, in the medium term, low prices are likely to delay the discovery and exploitation of some additional reserves, and to bring about the earlier abandonment of some existing (but relatively expensive) resources. This carries implications for availability – and hence for prices – in the longer term.

4.9 Putting this analysis together leads most people to envisage either broadly static or moderately rising oil prices over the next 10-20 years. An examination of some recent forecasts shows prices in the range $16-29 per barrel in 2000 and $16-32 in 2005, at 1995 prices. Paragraphs 4.16 – 4.25 consider the most recent set of DTI energy projections, published in March 1995 as Energy Paper 65 (*Energy Projections for the UK: Energy Use and Energy-Related Emissions of Carbon Dioxide in the UK, 1995 – 2020*). Two basic sets of assumptions are needed, about the rate of growth of the UK economy and about world energy prices. The DTI has had to make assumptions about possible long-term oil price trends in two previous exercises (published in 1989 and 1992), and the 1995 assumptions can be compared with these. The comparison is salutary (see Chart 4.1). As is apparent, expectations about the future course of oil prices have declined markedly. Expectations could, of course, change again. It is not impossible that major political upheavals could lead to sustained price hikes above the levels

Chart 4.1

**Crude oil price assumptions, 1990 to 2005
(Comparison of DTI assumptions)**

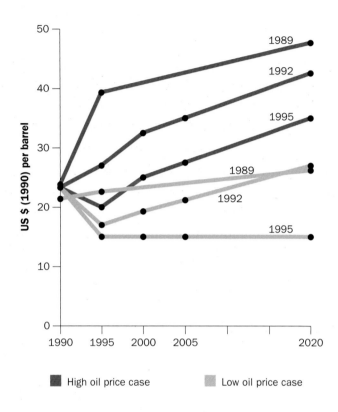

Source: Department of Trade and Industry

assumed here. But current assumptions are consistent with most people's expectations about world oil markets.

4.10 In the past, there has been some tendency for the prices of all fuels to be linked. But with the growth of world trade in coal, there has been a large measure of decoupling of coal and oil prices, with coal prices determined largely by the competition between coal producers, in a market with a tendency to over-supply. The long-term determinant of the price of coal is world marginal production costs. Ultimately these may rise, as demand grows and reserves are depleted, but current evidence is that they are more likely to remain low. There is plenty of coal to be had at around current prices: proved world reserves stand at around 200 years' supply at current production rates.

4.11 For gas, the picture may also be one of some decoupling from the price of oil. World gas markets are as yet relatively undeveloped, and there is still a set of essentially regional markets. Demand for gas is rising rapidly, but so too is supply, with proved reserves standing at 50 years' current production. Even so, gas supply in the medium to long term is subject to some uncertainty. The availability of supplies requires the necessary infrastructure for exploitation and transportation. Investment in such infrastructure will only take place if both producers and consumers believe that prices will cover these costs and still be competitive, both with other fuels and with other sources of gas. Perceptions about the political stability of producers, and of the countries crossed by the main transmission networks, play an important part in maintaining the confidence of the market. The traditional gas supply areas for Western Europe are Western Europe itself, Algeria, and Russia. The latter two are currently suffering political difficulties, and it remains to be seen what effect, if any, this will have on long-term confidence.

4.12 The DTI projections look at the likely course of UK gas prices. Here the assumption is that, as the UK becomes fully integrated with the West European gas transportation system, there could, in the medium term, be some tendency for beach prices to rise faster in real terms than oil prices and coal prices.

4.13 The broad picture for all fuels is of prices which, if they are to increase in real terms, are likely to rise fairly slowly. This does not preclude the possibility of sudden sharp price "spikes", but most people would expect to see these offset over time, such that prices come down again towards the longer term trend. The recent DTI projection exercise considered the effect of price hikes, possibly raising oil prices near to record levels, but it saw these as lasting no more than a few years, until the longer term influences on supply and demand had time to reassert themselves.

UK ENERGY PROJECTIONS

4.14 The UK's immediate energy future seems reasonably secure, given its indigenous resources and well-developed infrastructure of supply. The Government believes that market structures are in place which will provide security of supply in the long term, and it has no day-to-day need to predict the development of markets. Previous UK governments have issued the energy industries with a centrally-determined view of likely energy futures, as a foundation for the industries' own plans. This approach carries the very real danger of leading everybody in the wrong direction. In a market-based system, it is important that there should be a range of different ideas about the future. Some will prove better than others. The companies which prosper are likely to be those which either have been operating on the basis of the most accurate set of assumptions, or have recognised the uncertainties and planned robust strategies to deal with the risks.

4.15 The Government is not, therefore, in the business of providing a set of projections to be used as the basis for industries' plans. Nevertheless, it has its own reasons for wanting to develop some considered views of the future. Most obviously, the requirements of international environmental policy-making are such that governments have to be able to develop some ideas about the future demand and supply of energy as a basis for decisions about the commitments which they can make about future emissions of energy-related gases.

FUTURE UNCERTAINTIES

4.16 The Government's Climate Change Programme (see Chapter 5) took as its starting point the figures which were published by the Department of Trade and Industry in 1992 (Energy Paper 59, see Appendix 5). The DTI's model of the demand and supply of energy in the UK economy was used to produce a set of energy projections from 1995 to 2020. These were not forecasts, but alternative projections, using different assumptions about the trend rate of increase in GDP and about energy prices. The projections also required some assumptions about likely investment and other plant decisions in the electricity supply industry over the next few years.

4.17 Energy modelling has proved itself to be a very uncertain business. The DTI exercise produced a range of numbers representing this uncertainty. Given that one purpose of the exercise is to focus attention on future uncertainty, it is not plausible to attribute precise probabilities to each of the various outcomes. Although figures at the far extremes of the range are generally less likely than figures nearer to the middle, prudent policy-making should take into account the full range of outcomes, rather than concentrate on the centre of the range.

4.18 Like all projections of this kind, the figures are in part dependent on the underlying assumptions. Since the 1992 projections exercise, views have changed about some of the most appropriate background assumptions. At the same time, work has continued to improve the underlying structure of the DTI model. As a result, some of the results from the new projections in Energy Paper 65 (see paragraph 4.9) are significantly different to those published in 1992.

4.19 The biggest changes concern the projected use of different primary fuels in electricity generation, and in the longer term, estimates of the likely trend in energy demand from the various sectors – domestic, industrial, commercial, transport etc. Chart 4.2 (overleaf) shows how the picture has changed.

4.20 This comparison shows that there has been a reduction in expectations about the maximum plausible increase in both primary and final energy demand. This is explained in paragraph 4.31. In the period to 2000, the ranges shown in the latest projections are slightly narrower, but are not so very different from before.

4.21 The level of electricity demand obviously affects the demand for primary fuels used in generation. But the mix of primary fuels used in generation also affects the overall level of primary fuel use, because of the different conversion efficiencies associated with the different fuels. The largest recent change in the energy generation has been the rapid arrival of the combined cycle gas turbine (CCGT) as a source of electricity generation. This has the advantages over conventional coal-fired plant of being quick and cheap to build and of having a lower environmental impact because of its clean fuel and higher conversion efficiency.

Chart 4.2

Projected increases in primary and final energy demand

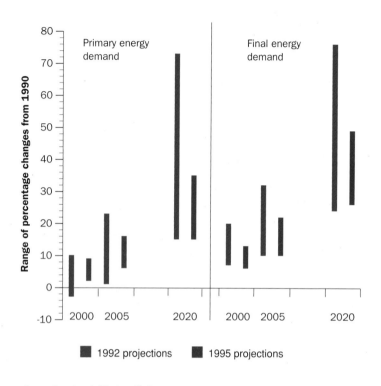

Source: Department of Trade and Industry

4.22 There is room for doubt about quite how dependent the electricity supply industry will wish to become on gas. Use of gas is expected to increase markedly, both in the UK and elsewhere in Europe, but there are risks in becoming too dependent on any one fuel. The switch from coal to gas has led to a more balanced and diverse range of fuels used in electricity generation. But given the markets' interest in maintaining a diverse portfolio of different fuel supplies, the growth in the use of gas could eventually slacken off. The issue is at what level this would happen. The DTI projections assume that generators' and consumers' concerns about maintaining a diverse generation mix will eventually limit the market share of any single fuel. This maximum share has been assumed to be no more than two-thirds of the electricity industry's generating capacity.

4.23 The projections contain some other points of interest about the possible future structure of electricity production. The results are driven by modelled estimates of the future cost of using different fuels. The share of traditional coal-fired capacity falls for the next 20 years, and at that point the number of traditional coal-fired stations reaches a low

4.23 level. The question is whether there are then any circumstances in which new clean technologies, which are being developed to burn gasified coal in CCGTs, could start to look economic. Not surprisingly, clean coal does seem to have a future if gas prices are assumed to be high.

4.24 The simulations also test for the potential for new nuclear build and suggest that, on the assumptions made, new nuclear stations would not be commercially viable. The Nuclear Review (see paragraphs 9.52- 9.54) will take a broader look at the potential role of nuclear power, and the EP65 conclusion does not prejudge the outcome of the Review.

4.25 The simulations show a continuing increase in the share of renewables capacity in total ESI capacity, rising from around 2½% currently to around 4% in 2000 and between 5% and 6½% in 2020.

TECHNOLOGICAL CHANGE

4.26 There are many studies of possible changes in energy technologies. Some assume a vastly different future, with major breakthroughs in technology leading to very different patterns of energy production and use, perhaps in response to stronger environmental regulation. There seems little doubt that in many areas, eg domestic and transport use, there is the technical potential for considerably reduced use of energy. The uncertainties lie in the cost of this technological improvement and how consumers and producers would respond to the new possibilities in the marketplace.

4.27 The DTI projections are based on the implicit assumption that the pace of technological change will continue broadly at current rates – certainly no discontinuity is envisaged. Other large-scale energy models adopt the same approach. Some contain underlying relationships which will "kick in" new technological possibilities when the price of conventional fuels reaches crisis levels (the so called "backstop technologies"). But this is not so very different in practice from the ability of the DTI model to move to new and renewable or nuclear sources of energy should the price be right.

4.28 As has been shown, the DTI projections do contain some increased use of renewable sources of energy, but this is not a major effect. Detailed studies of the prospects for new and renewable energy in the UK were published last year[1], and an annual update is due to published in the summer. Details of the Non-Fossil Fuel Obligations which support the development of renewable energy are given in chapter 9.

4.29 Looking more widely at the range of energy technologies beyond just new and renewable, the DTI published a general review in May 1994[2]. This appraisal examines some of the possibilities which could bring about considerable change in the way energy is made and used. It also

---
[1] " New and Renewable Energy: Future Prospects in the UK" – DTI Energy Paper Number 62; and ETSU's "An Assessment of Renewable Energy for the UK", both HMSO.
[2] "Energy technologies in the UK" (Energy Paper Number 61), HMSO.

assesses both the scale of the potential contribution each technology might make and the possible timescale for deployment.

4.30 A similarly broad view is being taken within the Energy Panel of the Office of Science and Technology's Technology Foresight Programme. This process seeks to establish networks between industrialists, engineers, and scientists. The Programme's Energy Panel (one of 15 Sector Panels) issued its report at the end of April 1995. The recommendations in this report were supported by a wide-ranging consultation process. Perceived market and technology opportunities in energy are identified for the UK over the next 10-20 years, with the objectives of wealth creation and improvement in the quality of life. Recommendations are made as to how these objectives can be progressed. The Energy Panel has looked at primary fuel supply, transmission, conversion, and end use. Sound economics, improved energy efficiency, and environmental concern have been central to its considerations. In drawing up its ideas for future markets, the Panel used a scenario approach, distinguishing between "business-as-usual", "shifting sands", and "green" and "non-green" possibilities. This work may feed back into the modelling of the likely pace and scale of the take-up of new energy technologies.

CHANGES IN DEMAND

4.31 An interesting feature of the new DTI projections is the way in which they tend to decouple GDP growth and energy demand in the long term. As before, the DTI model is based on an analysis of past reactions by both industry and consumers to changes in output, incomes, and energy prices. But the modellers have adopted a framework which incorporates some longer term constraints on demand, largely as a result both of overall ceilings on the likely use of energy in various end uses (like air conditioning or domestic refrigerators) and of technological change (e.g. the gradual improvement in energy efficiency of appliances, such as personal computers or printers). The new approach also makes allowance for such factors as higher building standards and further changes in the structure of manufacturing industry. This incorporation of so called "bottom up" factors has proved to be an important change.

4.32 There are, of course, many economic and social changes in prospect which could also impact on energy use, although their extent and direction are not yet clear. Some of the possibilities are listed below:

- increased "homeworking" or telecommuting by white-collar employees, possibly leading to significant savings in energy used to travel to work, though also potentially to increased energy use at home;

- similar possibilities for electronic shopping;

- an ageing population, which might consume proportionately more energy as heat;

- a high requirement for personal mobility;

- continuing innovation in the electronics industry, with further growth in the number of energy-consuming appliances in homes and offices, but with increased energy efficiency through innovation and electronic control.

4.33 Although explicit allowance for each of these possible changes has not been made in the DTI projections, at least some of their impact is implicitly reflected in the underlying model of energy demand and supply.

4.34 All developed nations have seen some decoupling of the growth of energy demand from the growth of GDP. This process is likely to continue. The energy ratio (see box on page 30) is expected in the DTI scenarios to fall by between 1.2% and 1.4% a year between 1990 and 2005, and by between 1.3% and 1.7% a year between 2005 and 2020. This compares with the actual fall of 1.7% a year between 1968 and 1990, but this was a period of very great structural change in the UK economy. The period 1990-1993 showed an increase in the energy ratio, ie energy use rose, while GDP fell.

Conclusion

4.35 The figures presented in the new DTI energy scenarios are subject to very great uncertainties. Some people see the prospect of even more fundamental shifts in technology. These technological possibilities are the source of some of the most interesting speculation about the future. But it is very difficult to predict what will happen. Experience shows that it is one thing to demonstrate a technology, but quite another to be certain whether or not it is likely to be adopted in the marketplace.

CHAPTER 5

ENERGY AND THE ENVIRONMENT

5 ENERGY AND THE ENVIRONMENT

SUSTAINABLE DEVELOPMENT

5.1 Sustainable development means meeting the needs of the present without compromising the ability of future generations to meet their own needs. This requires the reconciliation of the demand for economic growth and the need for environmental protection. Economic growth depends on the availability of secure and diverse supplies of energy: but all large scale energy industries have some adverse effects on the environment.

5.2 The discussion in Chapter 4 suggests that world energy supplies are at present plentiful. Estimates of primary fuel reserves are currently increasing faster than those fuels are being used up. One day, supplies of our current fuels will start to run down, but there is no knowing now what kind or quantities of energy the world will need then. In due course, market participants – at both the supply and demand ends – will start to anticipate future shortages. But today, given the ready supply of fuels, there is no need, purely from the perspective of availability, to reduce the use of energy. It is the wider environmental impacts of energy use which are most likely to be important in determining whether any given use of energy is consistent with sustainable development.

5.3 This chapter considers the environmental impacts associated with the use of energy. These range from the local to the regional and the global (see box). Greater detail is given in Appendix 2, which looks at seven major pollutants emitted from the energy industries, all of which can have a significant regional or global impact on the environment.

THE LINK TO COMPETITIVENESS

5.4 The twin objectives of competitiveness and sustainable development both have the same underlying aim – the maximisation of welfare and the proper use of resources. Over time, they can reinforce each other: policies which encourage sustainable development provide the right basis on which firms can build their position in the global marketplace.

5.5 Firms, and indeed nations, may be able to improve their competitiveness in the short run by ignoring environmental impacts. For example, regulations which enforce the cleanup of emissions tend to increase costs, and so can jeopardise the trading position of firms and industries, particularly if these standards are tighter than those being adopted elsewhere. Short-term conflicts between environmental objectives and the competitiveness of today's firms cannot be ignored. But it would generally be wrong for the UK to use its own domestic natural resources in a non-sustainable or environmentally-damaging way just because another country chose to use its resources like that.

THE ENVIRONMENTAL IMPACTS OF THE ENERGY INDUSTRIES

The energy industries have many interactions with the planetary systems that sustain human life: they tap environmental resources – coal, oil, and gas produced from living things in the remote past, or radioactive rocks within the Earth's crust, or the energy of the planet's rotation and gravitational field, expressed in hydro, wind or wave power, or the sun's energy that drives the whole living system and is trapped as heat, photoelectric energy, and in wood, charcoal and crops. The exploitation and use of these resources produces impacts on the environment.

Some impacts of energy industries and energy use are *localised*, for example as:

- urban and roadside air pollution, with particulates (black smoke), nitrogen oxides, carbon monoxide, volatile organic compounds, and lead from the diminishing volume of leaded petrol still in use;

- water pollution from power plants, industrial cooling systems, mine drainage, or onshore oil wells;

- marine pollution from accidental oil spills, illegal discharges from ships, or offshore drilling and oilfield operation;

- radiation and radioactivity from the fuel cycle wastes of nuclear power stations (although some such materials, released to atmosphere or ocean, also contribute to worldwide background radiation);

- taking of land for energy facilities such as power stations, wind farms, power transmission lines, coal mines, hydroelectric dams, and so on;

- disposal of wastes, such as gypsum from power station desulphurisation plant, furnace bottom ash, colliery spoil, or the contaminated equipment, redundant buildings, and wastes of the nuclear industry.

Other impacts are much broader. The following pollutants can have a significant *regional or global impact* on the environment.

- carbon dioxide ⎫
- methane   ⎬ greenhouse gases

- oxides of sulphur   ⎫ acid rain precursors
- oxides of nitrogen ⎬ (nitrous oxide is also
                      ⎭ a greenhouse gas)

- radionuclides.

5.6 The compatibility of the policies of sustainable development and competitiveness can be enhanced if the following principles are accepted when establishing environmental policies:-

- policies should take account, so far as possible, of comparisons of costs and benefits, including those social and environmental values which cannot readily be quantified in monetary terms;

- a range of policy instruments should be examined. Costs of compliance will be among the criteria for evaluation;

- full consideration should be given to the timing of policy implementation, so that the market has time to adjust: for example, it may be possible to reduce compliance costs by delaying implementation to allow improvements to be made during the normal capital replacement cycle.

5.7 Nevertheless, some commentators see a fundamental conflict between the objectives of environmental policy and the growth of competitive energy markets. They fear that where competition and economic efficiency are the drivers, environmental costs will be neglected or underestimated. This is to ignore the many positive links between economic efficiency and environmental protection. Specific examples of potential environmental benefits are:-

- commercial pressures in energy markets will tend to produce greater production and conversion efficiency;

- there will be commercial incentives to provide more value-added services, including efficient energy management;

- more transparent costing of infrastructure could lead to more decentralised production and better matching of electricity generation and consumer demand, leading to fewer large transmission lines and a reduction in the losses inherent in long-distance transmission (currently about 1.5%);

- innovation could lead to more efficient and less environmentally-intrusive methods of generation (eg local or "embedded" generation for a particular building or site; further development of photovoltaics; improved distribution by superconducting cables);

- tariff structures which better reflect marginal costs are likely to improve the use of existing generating capacity, by helping to trim demand peaks and reducing the need for additional 'peaking' plant. The effect would be to move demand from plants with typically low efficiencies, burning carbon-intensive fuels, towards more efficient plants running on cleaner fuels;

- public concern, coupled with higher road fuel prices, could promote the development of more energy-efficient forms of internal combustion engine for transport, reducing emissions per mile travelled.

ENERGY AND THE ENVIRONMENT

5.8 These are examples of the potential for competition in the energy sector to promote efficiency in generation and use, bringing about lower energy consumption and lower levels of emission to the environment. On the other hand, if competition reduces unit costs (and hence prices), it could lead to higher overall energy demand. Likewise, increased system efficiency, which moves demand around over the course of the day, could lower costs and prices, and so lead to increased demand. From the environmental standpoint, this might be a step in the wrong direction, but realistically it is unlikely, in the foreseeable future, to be a very large one. Indeed, it would be perverse to pursue an environmental policy of promoting energy efficiency by encouraging inefficiencies in energy production.

5.9 Some Government intervention in energy markets is still likely to be necessary, however. A requirement of sustainability is that those charged with industrial and economic policy-making accept the need, so far as possible, to internalise what are known as "environmental externalities" (that is, environmental costs which fall outside the normal processes of exchange in a market economy). Precise valuation of costs may always be difficult, particularly where the environmental impact is widespread (most obviously in the case of global climate change). But the aim should be to devise procedures to recognise external costs (such as damage to ecosystems, to health, or to buildings) in decisions about energy production and use. A nation which attempts to take them into account will tend to move in the direction of sustainable development.

5.10 The UK Government's policy is to prevent unacceptable environmental impacts by means of appropriate regulation of the activity of both energy industries and users and/or the use of economic instruments. The Government has a predisposition in favour of the latter, since they are potentially capable of delivering environmental objectives more efficiently. But the judgment on the most appropriate solution must be made on a case-by-case basis.

5.11 This chapter now turns to the impact of two specific environmental policies on the energy industries: policies on air quality, and, in particular, control of the emission of the precursors of acid rain; and policies on climate change.

AIR QUALITY

5.12 The UK's long term approach to air quality has been set out in a recent document[1]. The policies embodied there reflect all of the UK's commitments, including those under various EU directives. The Government has also signed and ratified the first Volatile Organic Compounds (VOC) Protocol under the United Nations Convention on Long-Range Transboundary Air Pollution (UN LRTAP). With a view to reducing

[1] Air Quality: Meeting the Challenge

concentrations of ground level ozone, which have tended to rise in recent decades, this commits participating countries to a 30% reduction in national emissions of VOCs by 1999, compared with 1988 levels. A national strategy has been published, and it is estimated that the UK will more than meet the 30% target by a combination of measures, including direct emission abatement, process changes, and product reformulation and substitution. The cost to the refining and other industries of meeting these new obligations is already considerable, but a joint DTI/Department of Environment "Best Practice" pilot study programme has been set up in order to identify the most cost-effective measures available in specific sectors.

5.13 Although road transport is the largest single source of emissions of oxides of nitrogen (NOx), which are of local and transboundary interest both in their own right and, together with VOCs, as precursors of ground-level ozone, large stationary sources are regulated by the EC Large Combustion Plant Directive (LCPD). This commits the UK to reductions in emissions from existing combustion plant larger than 50 MW of 60% by 2003 for sulphur dioxide (SO_2), and 30% for NOx by 1998, both compared with 1980. Limits are also set for particulate (smoke) emissions from new plant (controls on existing plant are such that power stations account for only about 5% of total UK emissions of black smoke). In addition, the UK signed the first Nitrogen Protocol under the UN LRTAP, which entails a commitment to return national emissions of NOx to 1987 levels by 1994, a level which the UK is on target to meet. National NOx emissions are expected to fall further as new, three-way catalyst equipped cars enter the fleet; as the programme of fitting low-NOx burners to all major coal-fired power stations and refinery plant is completed by 1998; and as gas-fired power generation increases its share of the electricity market. Negotiations have begun on a second UN Nitrogen Protocol, in which inclusion of emissions of ammonia from agriculture is under consideration.

5.14 While many of these plans affect the energy industries as a whole, the agreement to reduce sulphur dioxide (SO_2) has particular implications for the electricity industry. Sulphur dioxide emissions are a major cause of the increase in the natural acidity of rain that has probably contributed to the acidification of soils and freshwater in geologically-sensitive areas of Europe. It is thought that sulphate aerosol particles derived from these sulphur dioxide emissions may also, with stratospheric particles of volcanic origin, contribute some measure of "global cooling", by reflecting sunlight away from the Earth's surface. This may be offsetting some of the possible effects of increased greenhouse gas emissions, espccially in the northern hemisphere. Sulphur emissions are now falling: in Britain in 1992 they were just over 70% of the 1980 total. In June 1994, the UK Government signed the Second Sulphur Protocol under the UN LRTAP. This commits the UK to further reductions, such that emissions reach 30% of the 1980 level by 2005 and 20% of the 1980 level by 2010 (see paragraph 6.31).

5.15 The DTI's energy projections indicate that, even in the absence of further measures, substantial reductions in sulphur emissions are expected, in line with the figures set out in the existing UK Plan. This reduction, and the associated environmental benefits, will not, however, be achieved without cost, and there will be additional costs to be borne in meeting the UK's new commitment. Most estimates suggest that the electricity supply industry and the refining industry are likely to bear a significant part of these costs, though ultimately, of course, they will also fall on consumers.

CLIMATE CHANGE

5.16 One set of environmental problems associated with the use of energy is potentially so large that it could necessitate substantial changes in energy use in order to meet the requirements of sustainable development. This is the potential impact on the global climate of the accumulation in the atmosphere of carbon dioxide (CO_2) and other "greenhouse gases", including methane and nitrous oxide. CO_2 is an inescapable by-product of energy production which is based on any fuel containing carbon.

5.17 Many specialist reports have analysed the sources, pathways, fates, and impacts of greenhouse gases and other pollutants. For the past 5 years, the UN-administered Intergovernmental Panel on Climate Change (IPCC) has been assessing the likely magnitude of changes in world climate in different regions, their possible impacts, and the potential of different measures to ameliorate the problem. The IPCC's preliminary assessment inspired the UN Framework Convention on Climate Change, signed by over 150 States at the United Nations Conference on Environment and Development in Rio de Janeiro in June 1992. By the end of March 1995, there were 118 parties to the Convention, including the UK and other members of the European Union, together with the Union itself. Parties to the Convention have accepted the obligation to take measures aimed at returning their emissions of CO_2 and other greenhouse gases to their 1990 levels by the year 2000, as a first step towards achieving the ultimate objective of "stabilisation of greenhouse gases in the atmosphere at a level that would prevent dangerous anthropogenic interference with the climate system".

5.18 The UK published its programme of measures aimed at meeting this commitment in January 1994[2]. The UK's programme reflects a precautionary approach while the world's scientists continue their work to ascertain the extent of the threat. The UK accepts that the risk of climate change is sufficiently great to justify action now, even though scientific uncertainties remain. Such action includes

[2] "Climate Change: The UK Programme" Cm2427, HMSO

measures which bring other benefits, as well as greenhouse gas reductions, and measures which can be taken now at low cost to avoid possibly more expensive action later. Should the seriousness of the problem be confirmed, and should the world as a whole then commit itself to a programme of substantial abatement, the required reductions in CO_2 might well demand substantial changes in energy production and use.

5.19 The UK's programme is built on the expectation that similar action will be taken elsewhere in the world. In 1990 the UK's share of global energy-related CO_2 emissions was less than 3%, and this proportion is likely to decline considerably over the next decade as developing nations industrialise. The impact of UK domestic policies on global greenhouse emissions is likely to be small. While it is right that we, and the other states of the European Union, should take precautionary action, and demonstrate what can be done, our action must be set in the context of a global response under the UN Climate Change Convention.

5.20 The Government believes that the UK is on course to achieve its existing commitment to return emissions of CO_2 to 1990 levels by the year 2000. This is confirmed by the DTI's latest energy projections, which suggest that emissions in 2000 will be 6-13 million tonnes of carbon below 1990 levels. This reflects changes in energy markets as well as the impact of the UK Climate Change Programme. The growth of CCGTs, the improved productivity of nuclear generation, and the decline of coal have all played a major part. Table 5.1 summarises the latest figures and compares them with the earlier projections.

5.21 The Government is currently reviewing progress on the measures contained in the UK Climate Change Programme, and will

Table 5.1
DTI projections of UK CO_2 emissions

Million tonnes of carbon

	1992 projections	without CO_2 measures	1995 projections with full programme of CO_2 measures*
1990 (base)	158	158	158
2000	156-178	152-161	144-152
2005	165-200	na	154-165
2010	172-227	na	154-167
2020	186-283	na	173-193

Source: 1992: Energy Paper 59
         1995: Energy Paper 65
na – not available
* includes all the measures set out in the UK Climate Change Programme (Cm 2427), except that the fiscal measures have been altered in the light of the 1994 Budget.

publish an update as soon as possible. The programme includes measures aimed at improving energy efficiency, delivering economic as well as environmental benefits. The main changes to the original programme are that VAT on domestic fuel and power will not be raised above the current level of 8%, and the Energy Saving Trust is no longer expected to contribute carbon savings on the scale originally envisaged by the year 2000.

5.22 In developing future climate change policies, the Government will take account of increasing scientific understanding of the problems, the international response, and the relative cost-effectiveness of possible measures. At the first Conference of Parties to the Climate Change Convention, which took place in Berlin in March/April 1995, the UK Government was expected to call upon developed countries to agree to a new target to reduce all greenhouse gas emissions to between 5% and 10% below 1990 levels by the year 2010.

environmental reasons should form part of international policy, so that UK industry and commerce is not placed at a disadvantage with competitors in other countries.

Conclusions

5.23 Today's national and international debate about greenhouse gas emissions and climate change means that the conditions under which energy is produced and used may need to change in order to contribute to environmental sustainability. Due regard must be paid to the importance of energy and energy costs in the nation's international competitiveness. Any moves to increase the cost of energy to users for

CHAPTER
6

INTERNATIONAL ENERGY AND ENVIRONMENT INITIATIVES

6 INTERNATIONAL ENERGY AND ENVIRONMENT INITIATIVES

INTRODUCTION

6.1 The production and use of energy is circumscribed by a wide range of policies and initiatives, at both national and international level. Some policies and actions are aimed at facilitating exploitation, or promotion of energy markets; others are intended to deal with environmental concerns, current or potential, associated with energy. This chapter focuses on initiatives in the international arena.

EUROPEAN UNION

Energy Green Paper

6.2 In preparation for the 1996 Inter-Governmental Conference (IGC), the European Commission has reopened discussions on Community energy policy and, in particular, the need for an Energy Chapter incorporated into the Treaties. The subject was previously considered during preparations for the 1991 IGC, when a draft Energy Chapter (which was later withdrawn) was produced for inclusion in the Treaty on European Union .

6.3 Following discussions at the Energy Councils in May and November 1994, the Commission issued a Green Paper in January 1995. The Green Paper raises the issues of:

- the respective roles of the Community, Member States, and the private sector in pursuing European energy interests;

- security of supply;

- environmental considerations;

- international co-operation;

- R&D, including alternative energy sources and the nuclear perspective.

6.4 The UK Government welcomes this initiative, which it regards as a useful contribution to the debate on how to make European business more competitive and Europe more prosperous. The Commission and the UK Government have kept industry informed and involved in this debate. This process will continue throughout 1995.

Trans European Networks

6.5 Article 129b of the Treaty on European Union provides for the development of Trans-European Networks (TENs) for transport, telecommunications, and energy. On 8 February 1994, the Commission submitted proposals for two Council Decisions for the development of TENs in the electricity and gas sectors. The UK Government has been broadly supportive of proposals in these areas, because it believes that they would help stimulate the development of the internal energy market.

6.6 The first Decision lays down guidelines establishing objectives, priorities, and broad lines of measures, including the identification of 'projects of common interest' (PCIs). The second addresses the technical, administrative, legal, and financial obstacles to the development of TENs in the energy sector. These Decisions will enable the Commission to promote co-operation between network operators and Member States, and provide financial assistance to PCIs for feasibility studies and other studies aimed at improving technical co-operation.

6.7 Throughout the negotiations, the UK has successfully pressed for the guidelines to pay attention to the need to avoid distorting competition and to encourage the use of private sector funding for infrastructure projects.

6.8 At the November 1994 Energy Council, there was agreement in principle on the guidelines Decision, including a list of PCIs. However, in the absence of the European Parliament's Opinion, no formal decision could be taken. Formal agreement is expected during the first half of 1995. Discussions are continuing on the second proposal.

Liberalisation and the Single Purchaser Model

6.9 In December 1993, the Commission presented to the Council revised proposals for Directives to liberalise the gas and electricity markets. The main features are non-discriminatory licensing procedures for the construction of power stations, LNG facilities and grids; competitive tendering for the construction of new and replacement power stations and networks, as an alternative to licensing; unbundling of accounts; and, most significantly, "negotiated" third party access (TPA) to networks on the basis of commercial agreements. Discussions on the electricity proposal started in January 1994, and continued throughout the year. It was not possible to reach agreement, because of opposition from a number of Member States, led by France, to the concept of "negotiated" access to networks.

6.10 As an alternative to negotiated access to networks, France put forward the "single purchaser model", with the following features: the single purchaser would invite tenders for new capacity and conclude long term supply contracts; certain large consumers would be free to negotiate with suppliers in other Member States but contracts would be concluded by the single purchaser; generators would be able to export any surplus production; and a regulator would supervise the operation of the system. The Conclusions of the Energy Council on 29 November 1994 made it clear that the "single purchaser model" would only be incorporated in the final version of the Directive if it could be shown to be equivalent to "negotiated TPA" as regards access to the market, and to be compatible with the EEC Treaty. The Commission was asked to undertake a study on this.

Legal Action by the European Commission against five Member States

6.11 In July 1994, the Commission decided to initiate proceedings before the European Court of Justice against France, Ireland, Italy, the Netherlands, and Spain for maintaining monopolies over the import and export of energy.

Proposals for a Carbon/Energy tax

6.12 Although the proposed Directive on an EU-wide carbon/energy tax remains formally on the table, discussions have moved away from a mandatory tax. The UK government remains of the view that an EU-wide carbon/energy tax is not the most cost-effective or appropriate of the range of instruments available to return CO_2 emissions to 1990 levels by the year 2000, and continues to oppose the imposition of such a tax, both because of its effects on competitiveness and also because the UK believes that it is best left to individual Member States to decide how to meet their CO_2 reduction targets.

6.13 However, there have been two key recent developments. First, the German Presidency, in the second half of 1994, advanced moves towards an approach based on existing Directives setting EU minimum excise rates. Work in this area is expected to be taken forward in the wider context of the Commission's forthcoming biennial report on mineral oils.

6.14 Secondly, at Essen in December 1994, the Council of Ministers noted the Commission's intention of "submitting guidelines to enable every Member State to apply a CO_2/energy tax on the basis of common parameters if it so desires". This was an important departure from the previous discussions seeking a mandatory EU-wide tax. The December Environment Council subsequently endorsed the Essen conclusions advocating this optional approach.

Large Combustion Plant Directive

6.15 Under the terms of the Large Combustion Plant Directive (88/609/EEC) the Commission was required to report on Member States' implementation of the Directive by December 1994. The Commission is also required to review the Directive itself by June 1995.

Energy Technology Programmes

6.16 The EC's non-nuclear R&D programme (JOULE) and the THERMIE programme for the promotion of energy technologies both ended in 1994. These programmes are being merged into a new programme (JOULE-THERMIE) as part of the Fourth Framework Programme. This will provide support for both R&D and demonstration projects in the fields of rational use of energy, renewables, and fossil fuels. The new programme will last until the end of 1998, with a budget of around £800 million.

INTERNATIONAL ENERGY AND ENVIRONMENT INITIATIVES

6.17 The UK has been successful in obtaining funds from the JOULE and THERMIE programmes. 1994 was a particularly good year, with UK-led proposals winning more than 20% of THERMIE funds. The two biggest successful proposals involve the retrofitting of gas re-burn technology to Longannet power station, to reduce NOx emissions, and a project to build a 8 MWe power station fuelled with gasified biomass derived from short-rotation forestry.

Energy Efficiency Programme

6.18 The EC's "SAVE" (Specific Actions for Vigorous Energy Efficiency) programme was established in 1991, with a 35 MECU budget, to help achieve the aim of improving energy efficiency in the Community by 20% by 1995. A major part of the programme was the funding of two types of programmes: those needed to develop technical standards (100% funded), and programmes to support Member States' initiatives such as training and information programmes, developing networks, and sectoral programmes (30-50% funded). The UK has been successful in submitting suitable programmes.

6.19 The 1995 SAVE round is the last in the current scheme. The Commission is developing proposals for a further 5-year SAVE programme. The UK is concerned to ensure, amongst other things, that the management and monitoring of projects is strengthened, to ensure value for money and effective dissemination of results; and that the development of transnational networks and practical pilot projects is encouraged.

6.20 The Commission has introduced a number of specific regulatory measures to underpin the SAVE programme, including the non-traded goods Directive (93/79/EEC), and Directives on the labelling of fridges and freezers (92/75/EEC and 94/2/EC). Further Directives covering the labelling of other domestic appliances are in preparation, and a proposed Directive on minimum energy efficiency standards for fridges and freezers has recently been published.

Directive on Integrated Resource Planning

6.21 In February 1994, the Commission's Energy Directorate produced a first draft of a Directive on Integrated Resource Planning. This would seek to build energy savings into energy investment planning, and identified integrated resource planning as a means of achieving energy savings. To date, the draft has neither been formally agreed by the Commission nor submitted to the Council.

Voltage Harmonisation

6.22 The UK's Electricity Supply Regulations were amended in January 1995 to bring into effect an agreed change in the nominal voltage of the low voltage public electricity supply system in Great Britain from 240 volts to 230 volts, together with a change in the permitted supply variations, in accordance with the first stage of the CENELEC European voltage harmonisation proposals outlined in document HD 472-S1. The eventual harmonisation of the nominal supply voltage and permitted voltage variations throughout Europe will be of great benefit to manufacturers and consumers of electrical goods. Manufacturers will be able to offer single standard product ranges throughout Europe, which will remove a perceived barrier to trade. Consumers will have greater choice and flexibility in the purchase and use of electrical appliances.

INTERNATIONAL

Energy Charter Treaty

6.23 The negotiation of the Energy Charter has made an important contribution towards the reform of the energy sectors of Central and Eastern Europe and the countries of the Former Soviet Union (FSU). The UK has fully supported the development of the Charter since its initial proposal by the Dutch Prime Minister, Ruud Lubbers, in June 1990. The Charter was signed in The Hague on 17 December 1991 and subsequently by all OECD countries (except New Zealand) and all the countries of Central and Eastern Europe (CEE) and the FSU. Following signature, work commenced on the Energy Charter Treaty to translate the principles of the Charter into legally binding commitments.

6.24 The Treaty was signed by over 40 countries in Lisbon on 17 December 1994, and is the first:

- major economic agreement between the FSU and the West;

- international agreement enshrining free trade in energy products in international law;

- multilateral agreement covering investment and trade;

- agreement to elaborate rules of transit.

6.25 The Treaty aims to assist the restructuring of Central and Eastern Europe and the FSU by creating open, liberal, and non-discriminatory energy markets in those countries. Its objective is to protect foreign investors from political risks, thereby creating investment opportunities. The Treaty promotes the objective of market-oriented pricing, which will improve energy efficiency in those countries, thereby constraining pollution. Accompanying the Treaty is a Protocol on Energy Efficiency and Related Environmental Aspects, which specifically aims to promote, in the contracting countries, energy efficiency strategies and policies and legal and regulatory frameworks consistent with sustainable development, including the creation of conditions to further energy efficiency, with emphasis placed on cooperation between Contracting Parties. There are also provisions on transit, trade, competition, sovereignty over resources, and taxation.

INTERNATIONAL ENERGY AND ENVIRONMENT INITIATIVES

6.26 Negotiations will begin in 1995 on a second Treaty to enshrine the means of applying national treatment for the admission of investments.

International Energy Agency (IEA)

6.27 The early 1970s oil crisis led to the foundation of the IEA in 1974 as an autonomous forum within the framework of the OECD to implement an international energy programme. The IEA's objectives are to discuss and as far as possible to assist the co-ordination of its 23 member countries' energy policies, particularly action to be taken in the event of severe oil disruption. The IEA's main decision-making body is the Governing Board, composed of senior energy officials from each participating country. The Board also meets periodically at Ministerial level. It directs the activities of the Agency and makes the major policy decisions. It reviews the world energy situation, as well as domestic energy policies, to assess future energy supply and demand patterns and to determine policies to meet changing energy and economic conditions.

6.28 The UK strongly supports the work of the IEA. DTI provides the UK's Governing Board member and representatives on all IEA committees, with some support from the Foreign and Commonwealth Office (FCO). In June 1994, an IEA report found that the UK's energy policy was consistent with IEA's "Shared Goals" statement, which sets out a framework for future energy development. The report highlighted the benefits of increased competition in the electricity and gas sectors and wider choice for consumers. In December 1994, Robert Priddle, a former Deputy Secretary in the DTI, was appointed the new Executive Director of the IEA.

The United Nations

Climate Change

6.29 Over 150 countries, including the UK, have now signed the United Nations Framework Convention on Climate Change (FCCC), which came into force in March 1994. The UK was the first country to fulfil its commitments under the Convention and produce a detailed programme of measures aimed at returning greenhouse gas emissions to 1990 levels by the year 2000. Given the strong association of greenhouse gases (particularly carbon dioxide and methane) with the production and use of energy, the programme centred around measures to reduce energy consumption. The main features of the UK's Climate Change Programme were outlined in the 1994 Energy Report (see Appendix 5 for publication reference).

6.30 Meetings of the UN Intergovernmental Negotiating Committee (INC) on Climate Change, preparing for the first Conference of Parties to the Convention, were held in August 1994 and February 1995. The Conference took place in Berlin in

March/April 1995. At the time of writing, it was expected to review the existing national programmes and to examine the adequacy of existing commitments under the Convention, in the light of the latest scientific evidence on global warming. It was expected to be acknowledged that the existing commitments in the Convention were a first step, and unlikely to be adequate to enable the achievement of its ultimate objective, the "stabilization of greenhouse gas concentrations in the atmosphere at a level that would prevent dangerous anthropogenic interference with the climate system". Further action would be needed to limit greenhouse gas emissions beyond the year 2000, and work towards strengthening the Convention in this respect was expected. In this context, the UK was expected to call upon developed countries to agree to a new target to reduce all greenhouse gas emissions to between 5% and 10% below 1990 levels by the year 2010.

United Nations Economic Commission for Europe (UNECE)

6.31 In June 1994, Mr Gummer, Secretary of State for the Environment, signed on behalf of the UK Government the Second Sulphur Protocol under the 1972 United Nations Convention on Long-Range Transboundary Air Pollution. This Protocol commits the UK to achieving a reduction in total national emissions of sulphur dioxide of 80% by the year 2010, compared with 1980 emissions, with intermediate targets of 50% and 70% in the years 2000 and 2005 respectively. In 1994, the UK also ratified its earlier signature of the Volatile Organic Compounds (VOCs) Protocol, with its target of a 30% reduction in VOCs by 1999, compared with 1988.

6.32 Discussions have begun on a possible second protocol on emissions of oxides of nitrogen (NOx), under the above Convention. The first NOx protocol, to which the UK is a signatory, called for stabilisation by 1994 of emissions at the 1987 level, and the UK is expected to meet this target. Under discussion are the possible inclusion in a second nitrogen protocol of agricultural ammonia emissions as contributors to soil acidification, and of Volatile Organic Compounds (VOCs) as precursors (with NOx) of ground-level ozone. A first draft of this protocol is expected in 1996.

6.33 The UNECE Committee on Energy established the Energy Efficiency 2000 Project in 1991 as a result of the Ministerial Declaration of the Bergen Conference on Sustainable Energy Development in the ECE Region. The three-year project was renewed for a further three years in 1994. The main objective is to enhance trade and co-operation in energy-efficient, environmentally-sound technologies and management practices between participating States, in particular between formally centrally-planned economies and market economies. The project office is in

Geneva. Activities are financed through a trust fund or contributions "in kind" from participating countries.

6.34 The most interesting and practical idea to have emerged from discussions between East and West European delegates at the various seminars and meetings has been the development of "Energy Efficiency Demonstration Zones" in small areas of, for example, cities in eastern Europe, using western technology. The UK has encouraged industrial and commercial participation in EE2000 meetings, and a meeting on the financing of demonstration zones was jointly hosted by Rolls Royce Industrial Power Division and the Department of the Environment (DoE) in Newcastle upon Tyne in 1993. UK and Hungarian consultants and the DoE have produced manuals on "Financial Engineering for Energy Efficiency" for the EE2000 project. Apart from the technology transfer and commercial aspects of the work, the UK is seeking to ensure that UK participation and UN activities are cost-effective.

CONVENTION ON NUCLEAR SAFETY

6.35 In June 1994, following discussions in which the Government played an active part over a period of two years, an international Nuclear Safety Convention was agreed in Vienna. It will enter into force after ratification by 22 states, of which 17 must have at least one civil nuclear power plant. Fifty-five countries have now signed the Convention, and three have ratified. The UK plans to ratify during 1995. The Convention's aim is to achieve and maintain a high level of nuclear safety worldwide, by encouraging best practice in the safe design, construction, operation, and regulation of civil nuclear power plants. Each country ratifying the Convention will produce a national report stating how it meets or intends to meet a range of nuclear safety obligations set out in the Convention. These reports will be discussed at review meetings to be held at intervals of up to three years.

G7 NAPLES SUMMIT AND AGREEMENT ON ACTION PLAN FOR UKRAINE

6.36 Nuclear power makes a significant contribution to the energy requirements of many of the countries in Central and Eastern Europe (CEE) and the FSU, where the standards of safety are lower than in the West. Russia alone has 24 reactors, of which 15 belong to the two types considered to pose the highest risk, Chernobyl-type RBMKs and VVER 230s. The UK continues to play a major role in the international efforts to improve nuclear safety in the CEE/FSU. The G7 Munich Summit (1992) produced a multilateral action programme for urgent improvements to the higher-risk plants and led to the establishment of the Nuclear Safety Account at the European Bank for Reconstruction and Development (EBRD). The UK has made a contribution of £13.25 million to the Account, which has funded projects in Bulgaria and Lithuania, and in 1995 will launch major projects in Russia. The Tokyo Summit (1993) took the process a stage further by bringing in the International Finance Institutions (IFIs) and promoting the development of co-ordinated energy strategies in the CEE/FSU countries concerned, to allow the phasing out of the higher-risk reactors. The Naples Summit (1994) focused on one such strategy, for Ukraine, and pledged funds for implementation of an Action Plan for the whole energy sector, to incorporate early closure of the remaining units at Chernobyl. The Plan is under development by a joint Ukraine/Western task force, including strong representation by the IFIs and with regular progress reviews by the G7.

6.37 The UK's main contribution to this international effort is channelled through the European Union's PHARE and TACIS nuclear safety programmes, which concentrate on plant operational safety, design safety, and support to the regulatory authorities. Bilaterally, the UK is providing assistance through the Know How Fund and the DTI, as well as cost-free experts for programmes run by the International Atomic Energy Agency. The UK nuclear industry and the Nuclear Installations Inspectorate are also playing a major part in these programmes.

Nuclear Fusion

6.38 The Government continues to fund fusion R&D, which is carried out as part of the European fusion programme. In March 1995, the Joint European Torus (JET) Council approved a proposal to extend the life of the JET project by a further three years. A formal proposal to extend the project will have to be submitted to the Council of Ministers for its approval.

Part Two:
The Energy Industries

CHAPTER 7

THE ENERGY INDUSTRIES

7 THE ENERGY INDUSTRIES

7.1 In the last few years, the suppliers of energy have entered into a process of change from monopolies to more open competition. The following chapters set out developments in each of the energy industries.

> **THE ENERGY INDUSTRIES**
> - 5% of GDP
> - 10% of total investment
> - 42% of industrial investment
> - 10% of annual business expenditure on R&D
> - 3 of the "Top 10" UK companies
> - 300,000 people directly employed (5½% of industrial employees)
> - Many other employees depend on these industries indirectly (eg 250,000 in support of activity on the UK Continental Shelf)
> - Major contributors to the balance of payments (trade surplus in fuels of £3½ billion in 1994)

ENERGY COMPANIES IN INTERNATIONAL MARKETS.

7.2 The outlook of all the former public sector energy industries is becoming more international. The liberalisation of the energy sector, and the resultant competition from new entrants, has imposed increasing commercial pressures on the successor companies to the former state-owned energy industries. They have responded both by increasing the efficiency of their core businesses and by seeking to diversify, within the UK and overseas.

7.3 British Gas has used its experience as a vertically-integrated shoreline-to-customer gas supplier, and as a developer of offshore gas projects. National Power (NP) and PowerGen (PG) are designers and operators of large power facilities. National Grid Company (NGC) are experts in transmission systems, and the Regional Electricity Companies (RECs), are experts in distribution. This expertise, and that of the equivalent Scottish companies, has assisted the companies in securing a number of projects around the world.

7.4 The demand for energy, in particular in developing countries, is growing very rapidly. This is providing new opportunities for UK energy utilities which they would generally not have been able to pursue under the former regime. Greater market efficiency at home, and a new competitive presence abroad, increase the competitiveness of the UK economy in world markets.

7.5 The UK oil and gas operators are active overseas. BP, British Gas, and Shell have a wide portfolio covering all the oil and gas theatres of the world, while others have been more selective. There is a particularly strong UK presence in the FSU, the Pacific Rim, Latin America, the Middle East, and Europe.

7.6 The UK also has a broad range of skills and services to offer in the provision of renewable energy, from consultancy and equipment supply through to scheme development and operations.

THE ENERGY INDUSTRIES

> ### INTERNATIONAL ACTIVITIES
>
> The activities of energy companies in international markets are outlined in the various company annual reports. Below are some examples:
>
> - the annual report of the Electricity Association outlines various overseas ventures by the constituent parts of the electricity industry. There are examples of activities in China, the USA, and Portugal;
>
> - Nuclear Electric are bidding, with Westinghouse, to build a PWR in Taiwan;
>
> - the gas industry, and in particular British Gas, has involved itself in ventures in Latin America, Asia, and Europe;
>
> - new development opportunities for oil companies are apparent in many parts of the world which have previously been closed to foreign investment. These include several parts of the former Soviet Union. In the course of the past year, the Minister for Industry and Energy, Mr Tim Eggar, has led trade missions to Azerbaijan, Uzbeckistan, Kazakhstan, and Turkmenistan.

7.7 The process of change influences not only the suppliers of energy itself but also the related industries of energy capital goods, oil and gas equipment supplies, and energy efficiency equipment.

ENERGY CAPITAL GOODS

7.8 Changes in the structure of final energy markets are bound to impact on the demand for UK-produced capital goods - generating sets, transmission equipment, and so on. The traditional relationship between UK energy companies and UK capital goods suppliers was close. But even without privatisation, the procurement procedures of the EC Utilities Directive would have forced companies into more open tendering for business. Privatisation has tended to accentuate a development under which companies can expect to buy at keen prices from the best suppliers in the world (though often through UK subsidiaries).

7.9 The world market for energy equipment is fiercely competitive, with only a few major players competing globally. UK companies have formed international alliances to enable them to compete successfully.

7.10 The energy capital goods industry in the UK covers companies producing a whole range of energy products. It is difficult to put together comprehensive figures that cover the complete industry. But data are available for some of the main components of the industry (see Table 7.1). The figures show that for several categories of products, UK sales are worth between £1 billion and £2 billion. This compares to a contribution of about £30 billion from the energy industries to GDP. It is also worth noting that in several categories there is an international market for most of these goods. UK companies are extensive exporters, with exports accounting for a large proportion of sales.

Table 7.1
The energy capital goods industry in the UK

The Central Statistical Office (CSO) have recently published a new series of booklets containing annual PRODCOM product data. A total of 250 different product classes have been identified using the 1992 Standard Industrial Classification (SIC), and the CSO have published data on product sales, exports, and imports. The following products, including data for 1993, are relevant to the energy capital goods industry:

Heading	1993 Annual sales in the UK £ million
Central heating boilers and iron and steel radiators - excluding cast iron radiators	330
Steam generators, except central heating hot water boilers	666
Engines and turbines, except aircraft	806
Bearings, gears, gearing and driving elements	1,061
Furnaces and furnace burners	249
Non-domestic cooling and ventilation equipment	1,651
Electric motors, generators and transformers	1,800
Electricity distribution and control apparatus	2,082
Insulated wire and cable	1,479
Accumulators, primary cells and batteries	417
Lighting equipment and electric lamps	1,017
Electrical equipment for engines and vehicles *(nes)*	783
Other electrical equipment *(nes)*	1,584
Electronic valves and tubes and other electronic components	2,556
Radio and electronic capital goods	1,179

(nes = not elsewhere specified)

OIL AND GAS SUPPLIES

7.11 The supply of equipment to the oil and gas industry is a huge world market, estimated to be worth more than £100 billion a year. The DTI's Oil and Gas Projects and Supplies Office (OSO) helps companies to identify good markets for their products, taking account of the opportunities presented by the growing liberalisation of world markets. Latin America, in particular, offers considerable opportunities.

7.12 Increasing competitiveness is vital to the oil and gas industry if the UKCS is to sustain its attractiveness to investors in the face of continued low oil prices. The Cost Reduction Initiative for the New Era (CRINE) is reducing capital and operating expenditure by instilling a new culture on the UKCS. The initiative is industry-wide: projects are developed by integrated teams of operators, contractors, and suppliers working to a common objective and sharing both risk and reward. Savings of up to 30% have been achieved through such measures as making greater use of standardisation and functional specifications when ordering equipment. Oil companies are also co-operating to reduce costs by, for example, carrying out joint seismic surveys or sharing aircraft and ships. Government has also had a role to play: industry controls have been rationalised, and much unnecessary documentation done away with. In addition, DTI is contributing £100,000 to the CRINE secretariat's 1995 budget. CRINE has had a number of benefits. It has assisted British companies in securing major contracts on a number of projects currently in the construction phase (eg Hamilton/ Liverpool Bay, Conoco/Britannia, British Gas/Armada); it has facilitated the exploitation of smaller and more marginal discoveries, and helped to extend the life of some existing fields, thereby boosting Government revenues; and the new approaches and technologies are of great potential export value.

7.13 The oil and gas supplies industry is particularly strong in the fields of design engineering, project management, pipeline technology, subsea engineering, and construction and process equipment and materials. UK suppliers are active in most areas of the world, and are achieving some outstanding successes. A sector of growth in the world market is demand for equipment and vessels to be used for the transport of liquid natural gas (LNG).

7.14 A further area of significant growth in demand is FPSOs (Floating Production Storage and Off Loading Systems). In Autumn 1994, the Minister for Industry and Energy set up a working group of oil company representatives and OSO officials to address weaknesses in this area. The group reported earlier this year, with a number of recommendations covering vessel build, tanker conversions, managing contractor capability, small field requirements, impact on topside competitiveness of overseas builds, and financing requirements.
The recommendations were endorsed by Ministers, and are now being implemented by the industry and Government.

ENERGY EFFICIENCY SUPPLIES INDUSTRY

7.15 The efficiency with which energy is extracted or generated and delivered to the user (supply-side efficiency) is essentially a matter for the energy industries themselves. It is a natural part of their activity, and it is hard to define a distinct element concerned specifically with energy efficiency.

7.16 The efficiency with which energy is used (demand side efficiency) gives rise to more distinct activity. But most firms that promote energy efficiency do so as a by-product of some other activity. Companies manufacturing or installing insulation material, glazing, lighting, heating and air conditioning, and building management systems are generally engaged in a wide range of activities. This makes it difficult to define and monitor the component devoted to energy efficiency.

7.17 The energy efficiency industry also includes a range of service providers. These include energy efficiency consultants who give advice, and contract energy management companies who go one stage further and take over the running of and sometimes the risk associated with a company's energy needs. There are also activities which straddle the supply and demand sides such as combined heat and power (CHP — see paragraphs 9.64-9.67).

7.18 Several organisations represent parts of this fragmented industry. As part of the competitiveness initiative, the Government has been putting increasing emphasis on a partnership approach. It has also tried to encourage a more coherent approach from the industry.

7.19 There are few reliable statistics about the energy efficiency industry. This is partly a function of its disparate nature, which presents practical difficulties in defining appropriate categories and in collecting data. As part of the initiative to promote a more coherent partnership-based approach, the Government will discuss with the various industry representatives the practicalities of collecting appropriate data.

7.20 In 1994, the Association for the Conservation of Energy (ACE) produced a report, *Sparking off Efficiency*, which analysed various aspects of the market for energy conservation in buildings and noted the impact of various Government policies on the development of the industry. Elements of the industry have also been encouraged by market developments over the years. A major question for the future is the extent to which the increase in competition in the utility industries, and the way that competition develops, will affect the energy efficiency industry.

CHAPTER
8

COAL

8 COAL

THE STRUCTURE OF THE INDUSTRY IN 1994

8.1 At the end of 1993, British Coal (BC) had 22 operating collieries and employed around 15,000 mineworkers. By the end of 1994, when BC was sold to the private sector, these figures had fallen to 15 and 7,000 respectively. UK coal production fell from 68.2 million tonnes in 1993 to 48.0 million tonnes in 1994.

8.2 During 1994, BC was the dominant supplier in the UK coal market, producing 92% of total output, with the remainder coming from the licensed privately-owned sector, mainly from opencast mines. Even before privatisation, the picture was beginning to change. By the end of 1994, 9 mines which BC had previously closed, had been transferred under licence to new private sector operators. Recent mine closures have reduced deep mined production sharply, with production from opencast mines falling only slightly. Chart 8.1 illustrates the differing trends in opencast and deep mined production since 1970. In 1994, opencast mines operated by BC and around 50 private mining companies accounted for 35% of UK coal

Chart 8.1

Trends in coal production, 1970 to 1994

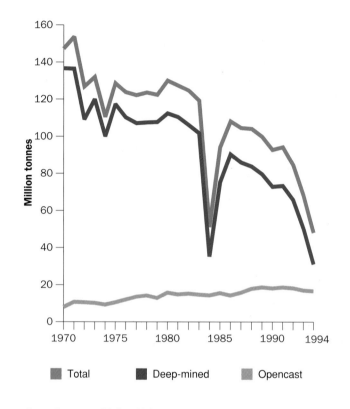

Source: Department of Trade and Industry

production, compared with only 12% in 1983. Detailed figures are shown in Table A10 in Appendix 1.

8.3 Between 1986 and 1994, labour productivity at BC's deep mines increased fourfold, and average operating costs fell by 57% in real terms. Over the three years to 1993/94, BC made operating profits after interest totalling £774 million. After allowing for

COAL

Chart 8.2
Trends in coal consumption, 1970 to 1994

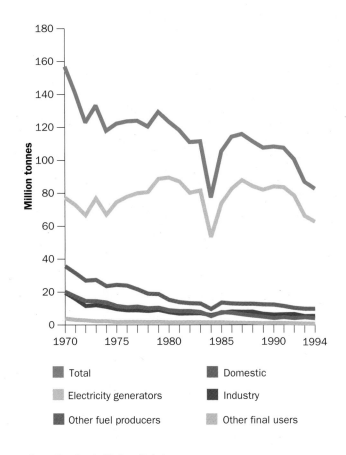

Source: Department of Trade and Industry

restructuring costs (e.g. pension contributions, redundancy payments, writedown of assets) a bottom line loss of £641 million was recorded over the period.

8.4 Deliveries of coal were 74.7 million tonnes in 1994 (down 14% on those of 1993). Trends in coal use are illustrated in Chart 8.2. (See Table A11 in Appendix 1 for more detail). Electricity generators continue to be the coal industry's dominant customer, despite a sharp decline in purchases since 1992. A comparison of Charts 8.1 and 8.2 shows that coal consumption has fallen less sharply than coal production since 1993. This difference largely reflects an increase in consumption of coal from stock over the period.

8.5 The rundown in stocks (see Chart 8.3) also partly explains the fall in coal imports since 1993 (see Table A10 in Appendix 1). Imports of steam coal, which is used mainly for electricity generation, fell from 8.4 million tonnes in 1993 to 6.1 million tonnes in 1994. Imports of other types of coal did not show this sharp decline: coking coal imports fell slightly, from 8.6 to 8.1 million tonnes, while imports of anthracite rose from 1.4 to 1.7 million tonnes. The UK produces relatively little of these categories of coal, which are used mostly by coke ovens and the domestic sector respectively. In both cases, therefore, demand must be met largely by imports.

INDUSTRIAL PRICES

8.6 In 1993, National Power and PowerGen and BC agreed 5-year contracts for the purchase of 40 million tonnes of coal in 1993/94, falling to 30 million tonnes in 1994/95 and each of the 3 subsequent years. BC also had contracts with the Scottish electricity generating industry for the supply of coal through to 1998. Coal prices are established in commercial negotiations between coal suppliers and purchasers. As part of its five-year contract with the two English generators, BC agreed to

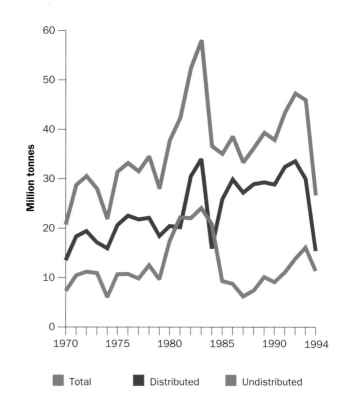

Chart 8.3

Trends in stocks of coal (at year end), 1970 to 1994

Source: Department of Trade and Industry

a pithead price of £1.51 per gigajoule (GJ) in 1993/94, the first year of the contract, falling to £1.33 per GJ in 1997/98, the final year of the contract (at October 1992 prices). The average price of power station coal imported into the European Union was about £0.98 per GJ in the first three quarters of 1994 (based on 24 GJ/tonne). This average reflects both contract and spot purchases, although most coal is purchased under contract. In 1994/95 the average spot price for a single delivery to North West Europe of imported coal was £0.89 per GJ. Trends in the spot price of coal in the main European markets since 1972 are illustrated in Chart 8.4.

PRIVATISATION

8.7 Legislation to enable privatisation of the industry to take place – the Coal Industry Act 1994 – was enacted on 5 July 1994.

Sale of British Coal's Mining Business

8.8 With the privatisation legislation making good progress through Parliament, the Department of Trade and Industry (DTI) pressed ahead with the preparations for the sale of five regional coal companies and seven care-and-maintenance collieries. Initially, in April 1994, potential bidders were supplied with a preliminary memorandum, on the basis of which they were invited to pre-qualify to participate in the bidding process. The DTI assessed each of the applications carefully before issuing a detailed information

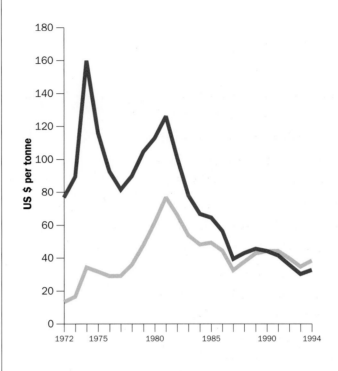

Chart 8.4
Trends in the spot price[1] of coal[2], 1972 to 1994

■ Price adjusted for inflation[3] (1990 prices) ▨ Current prices

(1) Amsterdam, Rotterdam and Antwerp.
(2) Imported steam coal mostly for power stations, adjusted to a common energy content of 26 GJ
(3) Calculated using the GDP deflator at market prices.

Source: Coal Week International, International Coal Report & Department of Trade and Industry

memorandum to successful pre-qualifiers and opening datarooms to provide the information pre-qualifiers required to formulate their bids.

8.9 By the closing date of 14 September 1994, 18 organisations had submitted bids for one or more of the regional coal companies and care-and-maintenance collieries. The DTI then selected preferred bidders for each of the companies. Following detailed negotiations with these selected companies, the sales of the relevant regional coal companies were completed at the end of December 1994, as follows:

- RJB Mining (UK) Ltd (English business and two care-and-maintenance collieries) to RJB Mining plc;

- Tower Colliery Ltd (care-and-maintenance colliery in South Wales) to a former employee buyout team;

- The Scottish Coal Company Limited to Mining (Scotland) Ltd;

- Celtic Energy Limited (South Wales business) to Celtic Group Holdings Limited (a management buyout team).

Total proceeds from the sales were some £960 million.

8.10 On the basis of the recent output of the various mines sold, RJB will have about 80% of deep mined output and about 70% of overall output.
Details of mine ownership are given in Appendix 3.

The Coal Authority

8.11 The establishment of the Coal Authority on 31 October 1994 was a major step in the privatisation of the coal industry. On that date, the Authority took up its full range of duties, powers, and functions under the Coal Industry Act 1994, and also took over ownership of Britain's coal reserves from BC. The Coal Authority's aim in carrying out its main tasks (see box) under the 1994 Act is to encourage viable coal mining and coal bed methane operations, whilst providing the required safeguards for the public against the effects of mining, particularly subsidence damage.

Licensing of closed pits

8.12 In the 1993 Coal Review White Paper, BC gave a commitment that any pits which it decided to close would be offered to the private sector for operation under licence. In total, 28 pits were offered for licence during 1993-94, of which 9 have transferred under licence to new private sector operators.

COAL

> ## THE COAL AUTHORITY
>
> The Authority has four main tasks:
>
> **to license coal mining:** the Authority will hold the nation's reserves of unworked coal previously held by British Coal, and will licence operators to work those reserves for mining or coal bed methane extraction. In doing so, the Authority will seek to secure the maintenance and development of an economically viable coal industry by licensed operators, that operators are able to finance the proper carrying on of mining operations and to meet the liabilities that arise from them, and also seek to ensure that operators have the funds to meet their subsidence obligations;
>
> **to take over British Coal's responsibility for certain property:** as owner of the unworked coal and underground workings formerly owned by British Coal, the Authority has also taken over British Coal's responsibility for dealing with events such as mine-shaft collapses, water discharges, or gas emissions arising from its ownership, except where these transfer to mine operators and new mine owners;
>
> **to provide information:** the Authority will continue to provide mining reports previously supplied by British Coal. It will also make publicly available geological data, mine plans, and information on subsidence, except where it is bound by confidentiality. The Authority will also keep registers of pending licence applications, of licences granted, of financial security arrangements for licensees' subsidence liabilities, and of orders made under the Opencast Coal Act 1958;
>
> **to deal with subsidence damage claims:** the Authority is responsible for dealing with subsidence damage claims, broadly outside areas of current mining. Subsidence damage claims within areas of current mining will be met by the licensed operators.

Non-operational property portfolio

8.13 British Coal announced on 11 January 1995 that it was to dispose of its non-mining property by way of separate portfolios for agricultural land, housing, and commercial development properties. The agricultural land and housing packages are generally being offered in a number of separate regional packages, while detailed decisions on the method of disposal of the other property have yet to be taken. It is BC's intention, as far as possible, to complete the disposal of all its property by Spring 1996.

Subsidiaries

8.14 Over the course of 1994/95, BC has been disposing of many of its non-mining activities, as part of the privatisation process. The main ones were:

- the sale of Compower Ltd to Phillips Communications and Processing Services Ltd;

- the outsourcing of BC's IT contract, also to Phillips in partnership with Origin UK;

- the sale of BC's shareholding in International Mining Consultants Ltd and Inter Continental Fuels Ltd to Coal Investment and RTZ;

- the sale of Coal Products Ltd to a management buy-out team;

- the sale of CRE Group Ltd to IMC Group Holdings*;

- the sale of TES Bretby Ltd, comprising businesses, assets, and staff transferred from the Scientific Services Division of the Technical Services Research Establishment, to John Mowlem Construction plc.

8.15 The principal non-mining activities remaining for the moment in the public sector are British Fuels Ltd, Centris, and CINMan, which are intended to be disposed of in 1995.

---

* The former Coal Research Establishment's large-scale R&D work is being retained within British Coal, which will continue to have some residual functions beyond 1995.

Consideration continues of the options for the future of British Coal Enterprise Ltd, the job creation arm of British Coal.

COAL MINING SUBSIDENCE

8.16 A large part of the Coal Industry Act 1994 was devoted to ensuring that after privatisation the rights of third parties, in particular in relation to subsidence, continue to be safeguarded. As well as bringing forward the essential features of earlier arrangements, it introduced additional safeguards for members of the public. These include a statutory right to arbitration, modelled on the lines of the earlier voluntary arrangements; and the appointment of a "Subsidence Adviser" who will not only provide advice, but can also investigate individual complaints.

8.17 The DTI has issued an updated version of its advisory leaflet for claimants, which contains all the necessary information to lodge a claim. These leaflets are widely available, and mine operators are obliged to send copies to people who may be affected by subsidence.

CHAPTER 9

ELECTRICITY

9 ELECTRICITY

INTRODUCTION

9.1 In **England and Wales** the electricity industry consists of companies concerned with:

- *generation*, i.e. the production of electricity: the main generators are National Power, PowerGen, and Nuclear Electric, but there is an increasing number of large independent power producers and a large number of small "autogenerators" who produce power mainly for their own use;

- *transmission and distribution:* the National Grid Company (NGC) owns and operates the transmission system, and is responsible for calling up generation plant to meet demand. Distribution to most customers is in the hands of the Regional Electricity Companies (RECs);

- *franchise supply:* the RECs have a monopoly of all franchise sales (i.e. sales to consumers taking less than 100 kilowatts) in their regions, and these RECs are known as first tier suppliers;

- *non-franchise supply:* above 100 kW, the market is open to competition, and consumers may have contracts with a second tier supplier, who might be one of the generators, a REC from another region, or an independent supplier.

9.2 The position in **Scotland** is different. ScottishPower and Scottish Hydro-Electric remain vertically-integrated companies which generate, transmit, distribute, and supply electricity to final consumers. (paragraphs 9.37-9.40 deal with events in the Scottish electricity market.)

9.3 In **Northern Ireland,** Northern Ireland Electricity plc (NIE) is responsible for power procurement, transmission, distribution, and supply. Generation is by other private sector companies (paragraphs 9.41-9.45 deal with events in the Northern Ireland electricity market).

GENERATION IN THE UK

9.4 The growth of gas-fired generation continues. In 1994, combined cycle gas turbines (CCGTs) accounted for 13% of the electricity generated by major power producers in the UK, compared to 8% in 1993. This increase was balanced by a fall in coal- and oil-fired generation.

9.5 The contribution of nuclear power in 1994 was broadly the same as in 1993 - about 27% of UK generation. However, this is expected to rise in 1995, following the connection of Sizewell B to the grid in February. During the winter of 1994/95, nuclear output was depressed by problems at 2 of the AGR stations, which led to the loss of over 2 GW of capacity during January, the period of highest electricity demand. Both these stations are now back in operation.

9.6 Generation by the use of new and renewable sources of energy is covered in paragraphs 9.55-9.61. The contribution of autogenerators and Combined Heat and Power plants is considered in paragraphs 9.62-9.67.

GENERATION IN ENGLAND AND WALES

9.7 In **England and Wales** 5 new CCGT stations, with a total capacity of 3.2 GW have begun producing electricity in the last 12 months, although some of them have experienced technical problems during commissioning. Three of these stations (capacity 2.0 GW) are owned by independent power producers (IPPs). In the same period, National Power and PowerGen have closed or mothballed some 6 GW of coal- and oil-fired capacity.

9.8 These changes have reduced National Power and PowerGen's combined share of capacity from about 71% in March 1994 to about 64% in March 1995. The share of the IPPs has risen from 6% to 9%, and that of nuclear from 15% to 18%, following the opening of Sizewell B. The remaining capacity comprises the interconnectors, pumped storage, and small autogenerators.

9.9 A further 5 CCGT stations, with a total capacity of 5.4 GW, are under construction; and another 6, with a capacity of 4.8 GW, have consents and connection agreements with NGC. Completion of all these stations would bring total CCGT capacity to over 19 GW, of which over 9 GW would be owned by IPPs, who are making a substantial investment in the newly structured industry.

9.10 National Power and PowerGen have announced their intention to close a further 2 GW of coal- and oil-fired capacity this year. Generators are obliged to give the Director General of Electricity Supply (DGES) six months' notice of their intention to close plant. The DGES may appoint an independent assessor, if he believes it appropriate, to evaluate whether the closure proposal is reasonable; the

THE POOL

Most electricity is traded through the *Pool*. Since it cannot easily be stored, demand has to be met by simultaneous production and consumption, and NGC balances supply and demand by calling on line generating plant as required on a half-hourly basis, the cheapest being called first. Generators are paid the *Pool Purchase Price*, which is the price paid to the most expensive generator running in that half-hourly slot *(System Marginal Price, SMP)*, plus a capacity payment. Suppliers pay the *Pool Selling Price*, which comprises the Pool Purchase Price plus a payment covering the costs of maintaining the stability of the Grid *(Uplift)*.

Pool prices can fluctuate markedly. Generators and suppliers protect themselves from this by *Contracts for Differences* (CfD), which are agreements to pay each other on a specified price basis for a fixed period. When the Pool price is higher than the agreed price, the generator pays the supplier the difference, and vice versa when the Pool price is lower.

financial implications of the closure for the company concerned; and the extent of outside interest in the plant.

9.11 Price movements in the Pool since vesting are illustrated in Chart 9.1. Table 9.1 summarises Pool price developments over the past two years.

9.12 Following significant Pool price increases in 1993, the DGES concluded that there was widespread concern at the perceived ability of National Power and PowerGen to raise prices at will, and considered whether to make a reference to the Monopolies and Mergers Commission (MMC). Subsequently he announced that he had received undertakings from these generators that they would use all reasonable endeavours over the period to the end of 1995 to negotiate the sale or disposal of 4,000 MW and 2,000 MW

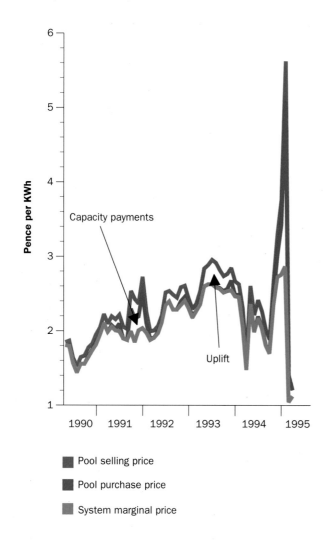

Chart 9.1
Electricity Pool Prices, 1990 to 1994

Source: Department of Trade and Industry (from data provided by Energy Settlements and Information Services Ltd).

ELECTRICITY

Table 9.1
Pool Purchase Price, 1993/94 and 1994/95

	p/unit	
	1993/94	1994/95
Weekdays	2.70	3.28
Weekends days	2.35	1.94
Night-time	2.07	1.26
Summer (Jun-Aug)	2.59	2.02
Winter (Dec-Feb)	2.37	3.30
Annual Averages		
time weighted	2.44	2.39
demand weighted	2.49	2.64

Source: Department of Trade and Industry (from data provided by Energy Settlements and Information Services Ltd).

respectively of oil-fired and coal-fired capacity. The aim is to promote competition by increasing the amount of independently-owned capacity in the system, and in particular to increase the amount of independently-owned capacity that operates away from baseload, since the new CCGT stations are generally running on baseload.

9.13 In late 1994 and early 1995, the DGES received reports from the companies on progress towards meeting the disposal undertakings. The DGES repeated in February 1995 that neither National Power not PowerGen had yet brought forward proposals which would meet the objective he had set out. He has encouraged both companies to explore a range of alternatives. If the disposals contemplated by the disposal undertakings are not completed by 31 December 1995, the DGES has said that he will consider the need for a reference to the MMC.

9.14 National Power and PowerGen also provided the DGES with an undertaking that they would bid into the Pool in such a way that, under reasonable assumptions of other generators' bids and taking seasonal fluctuations into account, average Pool Purchase Prices in 1994/95 and 1995/96 would in normal circumstances be expected not to exceed 2.4 p/unit time-weighted or 2.55 p/unit demand-weighted, at October 1993 prices*. The respective figures in 1994/95 prices are 2.46 and 2.61 p/unit.

9.15 Following these undertakings, Pool prices initially fell back. During the summer of 1994, the Pool Purchase Price averaged only about 2.0 p/unit, compared with about 2.5 p/unit the previous summer. However, from October onwards, prices rose increasingly quickly, reaching a record monthly average peak of 5.1 p/unit in January. Thereafter, prices fell sharply, to 1.4 p/unit in February and 1.2 p/unit in March. One of the reasons for the high prices in the early part of the winter was the temporary closure of 2 AGR stations (see paragraph 9.48). The loss of this capacity, at the time of year of highest demand, was one factor leading to a large increase in the capacity element of the Pool price which reflects the amount of spare capacity on the system.

---

* The time-weighted average is the average price over all the half-hour time slots in a year, obtained by adding up all the half-hourly prices and dividing by the number of slots. The demand-weighted average is obtained by multiplying the price and demand for each half-hour slot, adding together the figures for all the slots in the year, and dividing by total demand for the year. The impact of higher prices at times of high demand will produce a higher average than the simple arithmetical time-weighted figure.

9.16 Although time-weighted prices in 1994/95 fell within the limit of the undertaking between OFFER and the generators, demand-weighted prices exceeded the limit by about 1%. The DGES will consider what, if any, action he should take, taking into account the extent to which the breach of the demand-weighted price cap might be attributable to events that the generators could not reasonably have foreseen.

9.17 OFFER's February 1994 report noted that "Pool prices have been relatively flat since the Pool began" and that "the differential between peak and off-peak prices may well need to be greater than at present in order to sustain mid-merit and peaking plant in a competitive market". The 0.15 p/unit differential between the time-weighted and demand-weighted price caps reflected this concern. This year there has been a considerable increase in peak/off-peak Pool price differentials, both in daily and seasonal terms, as shown in Table 9.1.

9.18 Other Pool-related issues are:

- at the request of the Minister for Industry and Energy, the DGES investigated further the possibility of **trading outside the Pool,** and, after consultation, published a report in July 1994. The DGES acknowledged that trading outside the Pool could have potential benefits for those involved, but in view of the uncertainty of those benefits, the costs of developing the necessary arrangements, and the potential detriment to competition in the generation market, he concluded that a sufficient case had not been made for significant changes to existing arrangements. OFFER did agree, however, to reconsider this matter in future, and a number of players in the market, particularly large industrial consumers, continue to advocate the merits of such a scheme;

- in the same report, OFFER also considered the proposal that generators should be **paid according to the prices they bid,** instead of SMP. The conclusion was that risks for smaller generators would be increased without a strong likelihood of prices being lower, and that in the longer term, prices might even be higher. OFFER consequently rejected this option;

- the Pool works on the basis of bids by the generators to run their stations. But it is also possible to take bids from the demand side so that large consumers agree to cut their demand in return for compensation payments. A pilot **Demand Side Bidding (DSB)** scheme was introduced in England and Wales at the end of 1993. The scheme aimed to reduce demand by large customers at peak periods. The benefit to other consumers is that Pool prices at peak periods should be lower, offsetting the cost of meeting compensation payments. The benefit to the larger consumer is the compensation payment, with the balance of advantage depending on the value of that payment relative to the loss of output or other inconvenience. This scheme was initially scheduled to run to March 1994. It was extended by the Pool Executive Committee (PEC), first until June 1994 and then up to December 1994, to assist further evaluation. The PEC set up a Task Force, including representative customers, to assess the operation of the pilot scheme and to produce recommendations for more permanent and far-reaching arrangements. The pilot scheme never attracted more than about 10 participants, with a combined potential for load shedding of about 500 MW;

- a new **Uplift Management Incentive Scheme (UMIS)** was introduced in England and Wales in 1994/95 to give NGC a financial incentive to minimise Uplift (the component of the Pool price that mainly covers the costs of maintaining grid stability). So far, the scheme appears to be succeeding, in that the components of Uplift covered by the scheme have fallen by about 13% compared to last year. But UMIS does not cover capacity payments made to plant that is available but is out of merit. These "Unscheduled Availability" payments have risen compared to last year because of the general increase in capacity payments (see paragraph. 9.15). The UMIS will run to October 1995, and discussions are continuing about the best way of managing and charging Uplift costs in the longer term;

- further changes to existing arrangements are being considered by the Pool Executive Committee. Such changes include the removal of unnecessary barriers to local trades between generators and second tier suppliers, the adoption of a scheme designed to facilitate effective demand side participation, and the potential transfer of transmission-related elements of Uplift from the Pool to NGC.

9.19 Other developments concerning electricity generation in England and Wales are:

- to meet their targets for reducing sulphur dioxide emissions, the two major generators are fitting **Flue Gas Desulphurisation (FGD)** equipment to 6 GW of generating capacity. At the end of 1993, 1.3 GW of plant was equipped and operating with FGD. A further 1 GW of capacity was fitted with FGD at the end of 1994, and it is expected that FGD plant will be in operation with all 6 GW of capacity by the end of 1996. The two major generators have invested around £900 million in FGD. As they have reduced their sulphur dioxide emissions to 20% below their current targets, the two major generators do not expect to install further FGD equipment at coal-fired power stations. However, National Power's application for consent to modify Pembroke power station to burn emulsified hydrocarbon fuels includes the installation of FGD equipment;

- **orimulsion,** an emulsion of bitumen and water, is imported for use at two power stations. It is also being evaluated for use at a third power station, subject to its use being authorised by Her Majesty's Inspectorate of Pollution (HMIP). Orimulsion is a relatively cheap fuel and is likely to remain economically attractive, following a decision by the European Commission in April 1994 to reclassify it as a natural bitumen to which excise duty does not apply. However, its future potential as a fuel for electricity generation will depend on generating companies' commercial assessment of its viability, taking into account the measures HMIP may require from them to limit sulphur and particulate emissions;

- in March 1995, the Government disposed of its remaining shareholdings in National Power and PowerGen.

TRANSMISSION IN ENGLAND AND WALES

9.20 The National Grid Company (NGC) owns and operates the transmission system in England and Wales. NGC is responsible for the despatch of generating plant, and has a statutory duty to facilitate competition in generation and supply. NGC is owned jointly by the RECs, who each have different shareholdings according to their size. Recently the RECs have been looking at the possibility of disposing of their holdings in NGC, but no decision has been taken.

9.21 Generators and suppliers who use the National Grid are charged for their use of the system. In April 1993, the charging structure was changed so that it better reflects the costs of transmission. Cost-reflective charging differentials are being introduced progressively, and, in 1995/96, generators in the north of England will pay £7.76/kW, whereas those in the South West will **be paid** £6.49/kW. One effect is to encourage the construction of power stations in the south, where there is an imbalance of demand and generation. NGC also levies connection charges based on the value of assets employed. These charges are the subject of a current review which may result in changes from April 1997.

9.22 In 1992, OFFER asked NGC to review the security standards set out in NGC's transmission licence. Following NGC's August 1994 report, OFFER have concluded that:

- a limited relaxation of Operating Standards would deliver significant cost savings with little risk of increased unreliability;

- changes to Connection Standards would allow greater customer choice;

- there is no case at present for changes to the Planning Standards or Quality Criteria.

9.23 The grid in England and Wales is connected both to the Scottish grid and the French grid. Work has been completed on an upgrade of capacity on the Scotland/England interconnector from 850 MW to 1600 MW. This has enabled some increase in transfers. The full

1600 MW will, however, require the completion of the projected new transmission line through Cleveland and North Yorkshire. The application for consent to that line is still under consideration, and parts of the line are the subject of a current public enquiry. Power flows are predominantly from Scotland to England, and a further increase in capacity to 2200 MW is envisaged by the end of 1996, although this too will depend on the same new Cleveland/ North Yorkshire transmission line, as well as other reinforcement work, mainly to the Scottish transmission system.

9.24 The interconnector between England and France has a capacity of 2000 MW. Imports of electricity met about 5% of UK electricity requirements in 1994, about the same as in 1993.

DISTRIBUTION

9.25 The distribution systems consist of the wires taking electricity from the transmission system to customers' premises. Distribution charges account for about a quarter of the total cost of electricity to customers.

9.26 In August 1994, the DGES announced his proposals for new **distribution price controls in** England and Wales, which all 12 RECs subsequently accepted. In 1995/96, the first year of the new control, he proposed price reductions ranging from 11% to 17% in real terms, depending on the company concerned. For each of the following 4 years, prices were to be limited by the formula RPI-2%. The DGES estimated that these proposals would be worth between £70 and £90 to the average domestic customer over the next five years.

9.27 In December 1994, in anticipation of the expiry of the Government's special shares in the RECs, Trafalgar House made a hostile takeover bid for Northern Electric. The President of the Board of Trade, on the advice of the Director-General of Fair Trading, decided not to refer the proposed merger to the MMC, but sought voluntary undertakings in order to address regulatory concerns raised by the DGES.

9.28 From late 1994, there was growing concern that the DGES's distribution price control proposals would not be sufficiently challenging to the RECs, and would not represent a fair balance between the interests of shareholders and customers. The proposal documentation from Trafalgar House, along with Northern Electric's defence, added weight to these concerns. In March 1995, the DGES announced that he would review the position, and consider whether to tighten the controls further from 1 April 1996.

SUPPLY IN ENGLAND AND WALES

9.29 On 1 April 1994, the competitive market in electricity was extended to cover all premises with a maximum demand of more than 100 kW. Many of the 45,000 newly eligible consumers were subsequently able to achieve significant reductions in electricity prices. There were difficulties at some sites in installing and registering the necessary metering and information collection facilities, such that some bills were delayed. These problems are gradually being resolved by an Action Task Force set up by the Pool. Work is in hand to learn the lessons of the 1994 franchise drop, so as to avoid similar problems in 1998, when the remaining franchise is due to expire.

9.30 The change in market shares is shown by OFFER's annual survey of supply to the non-franchise market in England and Wales. The figures in Table 9.2 show market share both by number of sites and by volume. In the "over 1 MW" market, it is apparent that first tier RECs (ie those supplying in their own geographical region) are losing market share, while second tier RECs (ie those supplying outside their own region) are gaining. Other suppliers (including generating companies) have gained overall market share in terms of output and, over the period as a whole, have held steady in the number of sites supplied. In the newly-competitive "over 100 kW" market, the figures (which are estimates) suggest that the first tier RECs are so far holding on to about three-quarters of their former customers, with their principal competitors being second tier RECs rather than generating companies.

Table 9.2

Market Shares in Industrial Electricity Markets, England and Wales

	ABOVE 1 MW MARKET			100 kW to 1 MW MARKET
	1990/91	1992/93	1994/95 (estimate)	1994/95 (estimate)
SITES SUPPLIED				
First Tier RECs	72%	68%	55%	76%
Second Tier RECs	4%	12%	21%	19%
Others*	24%	20%	24%	5%
Number of sites	4,350	5,020	5,360	45,476
OUTPUT SUPPLIED				
First Tier RECs	57%	46%	37%	70%
Second Tier RECs	4%	12%	15%	21%
Others*	39%	42%	48%	9%
TWh supplied	75	76	74	38

* includes National Power, PowerGen, Scottish Hydro-Electric, ScottishPower, Nuclear Electric, and Independent Suppliers.
Source: OFFER

ELECTRICITY

9.31 A broad description of price trends has already been given in Chapter 2. This section gives more details about both the franchise and the non-franchise markets.

9.32 VAT at 8% was charged to domestic electricity bills from 1 April 1994. But there has been a continuing real fall in pre-VAT electricity prices paid by domestic consumers. Half the regional and national electricity companies reduced their domestic sector tariffs in April 1994, and half froze their tariffs at 1993 levels. Since April 1994, 4 RECs have announced rebates to domestic consumers, and 5 announced tariff cuts coming into force before 1 April 1995. In 1994 as a whole, prices to domestic users (before VAT) were some 4½% lower in real terms than in 1993. Over the last 5 years (between 1989 Q4 and 1994 Q4) domestic electricity prices have fallen by 3% in real terms (excluding VAT) but have risen by 4½% if VAT is included. However, even when VAT is included, prices to the domestic consumer are still lower than their peak in 1992.

9.33 Prices in 1995 are likely to be lower then in 1994. Six RECs announced tariff cuts to take effect from 1 April, with the remainder keeping tariffs unchanged. Including mid-year changes, 10 of the 12 RECs have now reduced tariffs from April 1994 levels, and when account is taken of inflation, all 12 have reduced tariffs in real terms. In same cases, further cuts have been made for particular customer groups, such as those paying by direct debit. Among other factors, these price cuts reflect both the continuing reduction in generation costs under the coal-backed contracts signed in 1993 and the impact of price controls.

9.34 Industrial electricity prices in the UK as a whole have fallen by about 11½% on average, in real terms, over the last 5 years, although within this average some companies have benefited from large real decreases, and some have incurred real increases.

9.35 Average industrial prices in 1994 were about 3½% lower in real terms than in 1993. One of the reasons for this reduction was the Pool price undertakings between OFFER and National Power and PowerGen (see paragraph 9.14), which was particularly important for some of the larger consumers who buy power at prices closely related to the Pool price.

9.36 However, Table 9.3 (overleaf) shows that the largest percentage reductions in 1994 were enjoyed by "Medium" size consumers, the group that benefited from the extension of the competitive market. A complete picture of the effect on prices for medium consumers of the opening up of the 100 kW market will not emerge until after the winter, but early indications are that this size of customer is achieving substantial reductions as a consequence of being able to switch suppliers.

Table 9.3

Trends in electricity prices paid by manufacturing industry, Great Britain, in real terms [1] [2] **(1990=100)**

	Size of consumer [3]				All manufacturing
	Small	Medium	Moderately Large	Extra Large	consumers
1989	95	108	111	103	107
1990	100	100	100	100	100
1991	103	96	93	102	97
1992	104	96	96	104	98
1993	95	96	98	108	99
1994	92	91	95	104	95
Average price level in 1994 (p/kWh) in out-turn prices	6.6	4.7	4.0	3.4	4.2

(1) Calculated using the GDP deflator at market prices
(2) These are average prices (exclusive of VAT) taken from an extensive survey of fuel prices paid by manufacturing industry in Great Britain, conducted quarterly by the DTI and the Central Statistical Office. These figures are based on the whole of the electricity bill, including standing charges and other fixed costs within the bill.
(3) Small consumers are those purchasing less than 880,000 kWh per year; medium consumers purchase between 880,000 and 8,800,000 kWh per year; moderately large consumers purchase between 8,800,000 and 15,000,000 kWh per year, and extra large consumers purchase over 150,000,000 kWh per year.

Source: DTI

THE ELECTRICITY MARKET IN SCOTLAND

9.37 The 2 authorised Public Electricity Suppliers in Scotland, ScottishPower and Scottish Hydro-Electric, are vertically-integrated companies, with activities covering generation, transmission, distribution, and supply. In addition to their own generating plants, the Scottish PESs purchase the entire output of the two nuclear power stations owned by Scottish Nuclear Ltd (SNL), under the Nuclear Energy Agreement put in place at vesting.

9.38 With generating capacity of about 10 GW against peak winter demand of under 6 GW, there is at present considerable over-capacity on the Scottish system. Some of the excess capacity is used to supply England and Wales, and exports are forecast to increase in line with the capacity of the interconnector (see paragraph 9.23) Most baseload demand in Scotland is met by the nuclear stations and the gas-fired station at Peterhead: additional demand is met by coal and hydro stations. In addition to plant owned by the 2 PESs and SNL, there are also about 25 small independent hydro stations, which sell their output to the PESs. That number is set to rise as more renewable energy projects are built in Scotland in response the first Scottish Renewables Obligation Order (see paragraph 9.55). ScottishPower are considering the installation of a seawater scrubbing FGD plant at Longannet.

9.39 As in England and Wales, there are regulatory constraints on the prices which may be charged for transmission, distribution, and supply in Scotland. The feature of regulation in Scotland is a direct control on allowed generation costs that can be passed through to Scottish franchise customers. In September 1994, the DGES proposed a tightening of the existing price controls on distribution to RPI-2 for ScottishPower and RPI-1 for Scottish Hydro-Electric. On supply, he proposed a control of RPI-2 for both companies. He also proposed to introduce competition in to the provision of connections to a PES's distribution system. Scottish Hydro-Electric declined to accept these proposals, and the matter has been referred to the MMC.

9.40 NIE plc and ScottishPower plc signed Heads of Agreement in September 1991 for the supply of electricity through a Northern Ireland (NI)/Scotland electricity interconnector, to be owned and built by NIE. The capacity of the interconnector will be 250 MW, and will meet about 20% of NI's future power requirements. The project is costed at about £200 million, and requires the extension of the overhead high voltage transmission system in Scotland from Coylton (south of Kilmarnock) to Currarie Port (in Ayrshire) and the laying of an undersea cable from Currarie Port to the Antrim Coast in Northern Ireland. Following opposition from residents to the overhead line proposal, a public inquiry opened in October 1994 to determine the issue of planning permission. A public inquiry into the Northern Ireland element of the project began in January 1995. The DGES (NI) announced in November 1994 that he would be undertaking an investigation into NIE's decision to proceed with interconnection to Scotland. The Government will reach a final decision on the project once the reports of the public inquiries have been considered.

THE ELECTRICITY MARKET IN NORTHERN IRELAND

9.41 The electricity system in Northern Ireland is small, heavily oil-dependent, and not interconnected to any other system. Responsibility for the transmission, distribution, and supply of electricity lies with Northern Ireland Electricity plc (NIE), which was privatised in 1993. There are three private companies generating electricity from four power stations. The largest power station, which is owned by British Gas, is being converted from oil to gas firing and accounts for almost half Northern Ireland's generating capacity. Total capacity is 2243 MW.

9.42 The small scale and isolation of the Northern Ireland electricity system make it difficult to introduce full scale competition quickly. In December 1993, OFFER (NI) published consultation proposals for the introduction of competition. It subsequently engaged Coopers and Lybrand Ltd to produce detailed proposals for a fully competitive electricity trading system in Northern Ireland. A consultative paper was published in December 1994.

9.43 Large users have complained that electricity prices in Northern Ireland are higher than those in Great Britain, and that the price differential places them at a competitive disadvantage. In June 1994, OFFER (NI) published a survey of electricity prices paid by large users in Northern Ireland and Great Britain in 1993/94. The broad conclusion was that the difference in average prices was of the order of 5-10%, but that this average hid a wide range of experience.

9.44 The proposal for a Scotland/Northern Ireland interconnector is dealt with in paragraph 9.40. This would end the current isolation of the Northern Ireland system, and would also help to meet NIE's additional power requirements in the latter half of this decade and beyond.

9.45 In November 1994, NIE and the Irish Republic's Electricity Supply Board announced their decision to restore the main Northern Ireland/Republic of Ireland electricity interconnector, which was destroyed by terrorist action in 1975. The restored link is expected to be operational by Spring 1995, and will provide mutual support during temporary shortfalls on either system, reducing the need for spinning reserve (plant which is up and running, not actually producing electricity, but ready to do so quickly if demand suddenly surges).

The interconnection will lead eventually to the linking of the two systems into Great Britain and wider European electricity networks.

NUCLEAR ELECTRICITY

9.46 In order to ensure diversity of electricity generating capacity in England and Wales, the Government has used powers under the Electricity Act 1989 to place a Non Fossil Fuel Obligation (NFFO) on the RECs, requiring them to contract for specified amounts of electricity from non-fossil sources. This is done by Statutory Instruments or Orders. The RECs contract for this non-fossil capacity at above-market prices, with the difference being reimbursed to them via the Fossil Fuel Levy (currently 10%) on electricity sales. Most of the NFFO premium-priced electricity is nuclear-sourced.

9.47 Under its NFFO contract, Nuclear Electric (NE) is paid a premium price on output up to a certain limit, in each year until the contract ends in 1998. For output above this level, NE gets the same price as other generators. NE expects these prices to enable the full costs of nuclear generation to be met, including the "back end" costs of decommissioning existing power stations. NE has closed its Magnox stations at Berkeley (1989) and Trawsfynydd (1993), and decommissioning is proceeding.

9.48 In 1994, nuclear stations in the UK produced about 80 TWh of electricity, accounting for about a quarter of the total electricity available. Production was about 1 TWh less than in 1993, partly because of problems with two AGR stations. Dungeness B was unable to come back online in October after its routine maintenance and inspection outage, and returned to operation in March 1995. Similar problems affected the Heysham 1 station, which was out of action during January 1995.

9.49 The 1.2 GW Pressurised Water Reactor Sizewell B is an important addition to UK nuclear capacity. Authorisations for the disposal of radioactive wastes required to enable Sizewell B to come into operation were granted by HMIP and the Ministry of Agriculture, Fisheries and Food (MAFF) on 2 November 1994, and came into operation on 2 December 1994. The station started producing power on 31 January 1995, and was expected to be operating at full capacity by April 1995. The cost of the station was some £2.5 billion (out-turn prices), which was within the target set by Nuclear Electric in 1990. The project was also completed broadly in line with the timetable set in that year.

9.50 In Scotland, the Nuclear Energy Agreement commits ScottishPower and Hydro-Electric to take all of Scottish Nuclear's output until 2005, and commits the latter not to sell electricity to third parties. The purchase price was fixed until 1993: after 1998, it will be set in relation to the market price for baseload generation in England and Wales, and between 1994 and 1998 it will taper down towards that price.

9.51 For information on decommissioning and radioactive waste (NIREX and BNF), see Appendix 4.

9.52 The start of the Government's Nuclear Review was announced in the first Energy Report. The Government believes that nuclear power's future role in the UK's electricity supply will depend on it proving itself competitive while maintaining rigorous standards of safety and environmental protection.

9.53 The Coal Review established that there is a sound economic basis for continuing to operate existing nuclear stations. The Nuclear Review is examining the economic and commercial viability of new nuclear power stations in the UK, and considering, without commitment, privatisation. It is also considering whether new nuclear power stations offer particular benefits or advantages in terms of the environment and of diversity and security of supply. The Review is also assessing the existing arrangements to enable the full costs of nuclear power to be met, and considering how best to manage the industry's waste and decommissioning liabilities, which are currently the responsibility of the public sector.

9.54 A public consultation period followed the announcement of the Review in May 1994. The industry published its case early in that period, to allow an informed debate, and over 500 submissions were received in total, some 200 of which were substantive responses. At the time of going to press, the views put forward are under consideration, together with advice from BZW, KPMG, and NERA, who are looking at privatisation/private finance, liabilities management, and diversity issues respectively. There is no date set for announcing the conclusions of the Review, although an announcement will be made as soon as is practicable once consideration is completed.

NEW AND RENEWABLE SOURCES OF ENERGY

9.55 The Government's programme aims to promote the development of new and renewable technologies and of the associated industrial and market infrastructure, so that they are given the opportunity to compete equitably in a self-sustaining market. The programme is described fully in Energy Paper 62, *New and Renewable Energy: Future Prospects in the UK,* published in 1994. This develops the themes in the Government's Competitiveness White Paper – particularly the encouragement of internationally competitive industries, by identifying and disseminating best practice information, encouraging companies to improve their skills base, and providing access to a range of business services. The programme has two main strands: the first involves stimulating an initial market for electricity-producing technologies close to commercial competitiveness via the Non Fossil Fuel Obligation (NFFO) in England, Wales, and Northern Ireland, and via the Scottish Renewables Obligation (SRO); the second is a supporting programme of assessment and market enablement, with resources being concentrated on those technologies with good prospects of commercial application.

9.56 The NFFO/SRO is a competitive scheme. The purpose of NFFO/SRO Orders is to create an initial market, so that in the not-too-distant future, the most promising renewables will be able to compete without financial support. This will require there to be a steady convergence under successive Orders between the price paid under the NFFO/SRO and the market price. This will only be achieved if there is effective competition in the allocation of NFFO/SRO contracts. Only those technologies identified as likely to be contributing economically to the UK electricity supply by the year 2005 are given market support via the NFFO/SRO. These currently include biofuels, wind, and hydro.

9.57 Three Orders have been made for England and Wales, the most recent in December 1994. Scottish and Northern Ireland Renewables Orders were also made in 1994. A breakdown of capacity of the main technologies involved in these recent Orders is shown in Chart 9.2. Second Orders for Northern Ireland and for Scotland, and a fourth Order for England and Wales, are in prospect for 1996.

Chart 9.2
Technologies supported under recent non-fossil fuel and renewables obligations, UK[1]

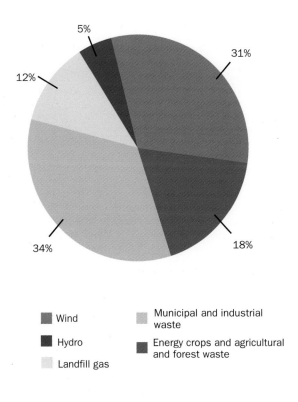

- Wind
- Hydro
- Landfill gas
- Municipal and industrial waste
- Energy crops and agricultural and forest waste

(1) Contracted capacity (MW DNC) under the Third Order for England and Wales and First Orders for Scotland and Northern Ireland.

Source: Department of Trade and Industy

9.58 As a consequence of the NFFO/SRO, the proportion of electricity generated from renewables (other than large-scale hydro) has increased rapidly from about 0.8 TWh in 1990 to 1.7 TWh in 1993, with a further increase expected in 1994. At the end of December 1994, 144 projects contracted under the first two Orders were in operation, with declared net capacity of 326 MW.

9.59 A report by the National Audit Office (NAO) on the Government's renewable energy programme was published in January 1994. Its broad conclusions were that:

- the DTI's methodology for assessing different renewables was soundly based, and a useful starting point in determining the priorities for funding;

- the influence of a few major customers, particularly the nationalised electricity industry, had led to the balance of the programme being tilted to a few large projects and programmes. There had not been sufficient support for projects aimed at the export market.

There was no suggestion of a lack either of value for money or of control of expenditure.

9.60 The DTI's updated strategy in EP 62 took full account of the NAO conclusions.

9.61 The Public Accounts Committee subsequently considered the NAO report, and published its own report in September 1994. It did not make any recommendations, although it did say that it doubted that the relatively modest increase in new electricity generation justified the sums spent on renewables. The Government believes that very good value for money has been obtained, both in absolute terms and in comparison with other national programmes. There is a large potential for these energy sources to contribute to sustainable energy supplies in the future, and they are already helping to reduce greenhouse gas emissions. Without the programme, there would have been little industrial interest or infrastructure, and the programme has been the key to unlocking the future contribution of renewables.

NEW AND RENEWABLES

Renewable energy is energy which occurs naturally and repeatedly in the environment and which can be harnessed for human benefit. The main carriers of the energy are the wind, the oceans, crops, the fall of water from lakes and rivers, and animal and human waste. The Government's policy is to stimulate the development of new and renewable energy sources wherever they have prospects of being economically attractive and environmentally acceptable in order to contribute to:

- diverse, secure and sustainable energy supplies;
- reduction in the emission of pollutants;
- encouragement of internationally competitive industries.

In doing this, it will take account of those factors which influence business competitiveness, and will work towards 1,500 MW of new electricity generating capacity from renewable sources for the UK by 2000.
The main renewable technologies identified as relevant to the UK are Biofuels (landfill gas, wastes, energy crops), Solar Energy (active, passive, and photovoltaics), Wind, and Hydro. Among the new energy sources under current investigation are Advanced Fuel Cells.

New and renewable energy sources are also an essential element of the Government's approach to reduce environmental damage caused by emissions of polluting gases. Achieving 1,500 MW of renewable energy capacity in the UK by 2000 will achieve annual savings of carbon dioxide amounting to 2 million tonnes of carbon, together with savings of 100,000 tonnes of SO_2 and about 30,000 tonnes of NOx.

AUTOGENERATORS AND CHP

9.62 There is a large number of small generators, mainly supplying their own needs, but in some cases also selling excess supplies to other users.

9.63 In May and July 1994, the Government introduced certain amendments to the Exemption Order (which sets out the categories of person who may generate or supply electricity without a licence). These deregulatory measures allow more small generators to operate without a generation licence, permit the resale of small amounts of electricity without the need for a supply licence, and permit more on-site generators to operate without a supply licence. In March 1995, the Government introduced a further amendment to this Order, which will extend to March 1998 the transitional period under which certain companies - who generated or supplied electricity before the electricity industry was vested - are exempt from the need for a licence.

9.64 Combined heat and power (CHP) plants are specially-designed energy systems which produce both electricity and usable heat. By extracting the benefit of the heat produced in the generating process, rather than shedding it (as in conventional generation), they can convert over 80% of the fuel input into useful output. As part of the UK's Climate Change Programme, the Government has set a target of 5,000 MW of installed CHP by the year 2000.

9.65 Growth in CHP capacity and generation over the 5 years from 1988 to 1993 was about 10% a year on average, and the proportion of UK electricity generation accounted for by CHP grew from 3% to $4\frac{1}{2}$%. There is evidence that the level of orders dropped during 1994. Opinions differ as to the reasons for this, and the issue is complicated by the long lead time for such investment. OFFER is required to establish and keep a database for CHP: by February 1995, it estimated total capacity at 3,300 MW. This figure, combined with evidence of orders picking up in early 1995, suggests that capacity is broadly on track for the target for the year 2000.

9.66 There are now over 1,100 CHP sites in the UK. The majority of these are in the services and residential sectors, particularly in hotels, leisure centres, and hospitals, where there is a high heat load and round-the-clock demand to make such systems attractive: but because the majority of such schemes tend to be small, over 90% of CHP capacity is to be found in the industrial sector, especially in the chemicals, oil refining, electricity generation, food, and paper-making industries.

9.67 CHP has benefited from the disaggregation of the electricity industry and from its liberalisation. The Government has taken various steps to amend regulations and licence requirements, which should help the development of CHP. The Government also assists in promotion of CHP through the Best Practice Programme (see paragraphs 3.24-3.26). Residential applications of CHP have been encouraged by a grant scheme organised by the Energy Saving Trust and financed by British Gas, with the approval of OFGAS. OFFER has recently announced that certain CHP schemes will be eligible for support under the Standards of Performance arrangements established with the electricity companies.

CHAPTER
10

GAS SUPPLY

10 Gas Supply

10.1 In 1994, there were further important developments in both the size and structure of the UK gas market. Prices continued to fall in most sectors, both in absolute terms and relative to prices paid by our main European competitors. The phased introduction of competition at home, and the slow beginnings of liberalisation on mainland Europe, mean that gas will remain a vibrant area of the energy market for at least the rest of this decade.

UK MARKET DEVELOPMENTS

10.2 Following growth of 12% in 1993, UK gas demand grew by 6% in 1994, a figure which would have been higher – at around 8% – but for an exceptionally mild November and December, which markedly reduced household demand. Table A16 in Appendix 1 gives more detail, including demand by sector over recent years.

10.3 Chart 10.1 gives a breakdown of gas consumption by sector for the period 1990 to 1994. By far the fastest-growing component was the power-station market. During the year, 4 major power-stations (Deeside, Keadby, Little Barford, and Barking) began producing electricity. These 4, together with the full-year contribution from the 7 stations that came into operation in 1993, account for a 45% increase in gas used for electricity generation in 1994, compared with 1993.

Chart 10.1
Consumption of gas[1], 1990 to 1994

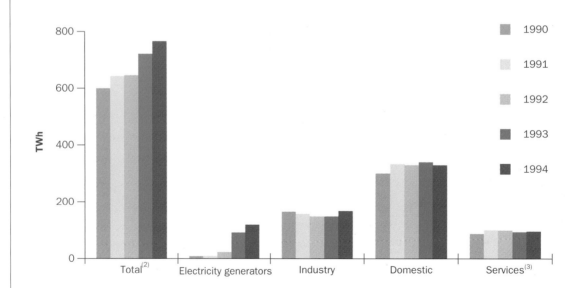

(1) Includes natural gas from North Sea and other UKCS fields, colliery methane, landfill gases and sewage sludge gas.
(2) Includes consumption by fuel producers (including electricity generators).
(3) Public administration, commerce and agriculture.

Source: Department of Trade and Industy

10.4 Demand for gas is expected to grow in all sectors. In the electricity sector, a further 4 major gas-fired power stations, with a combined capacity of over 2,000 MW, are expected to come into commission during 1995. Investment in CHP, environmental constraints on fuel oil use, and the recovery in economic activity are together expected to lift industrial and commercial demand. In the household market, demand will remain very sensitive to temperature. But underlying demand is continuing to grow, albeit more slowly than in the past, as the gas network is extended to rural areas and as gas-fired central heating is installed in more households. These trends are likely to be offset by more cost-reflective charging for new connections, by increasing saturation of the central heating market, and by the introduction of VAT on domestic gas.

10.5 An increase in UK demand is likely to follow the construction of a natural gas pipeline from Portpatrick in Scotland to Northern Ireland, with completion expected by the end of 1996. The primary market is Ballylumford station, which was purchased by British Gas in 1992 and is being converted from oil to gas firing; but the onshore line will be extended to serve wider industrial, commercial, and domestic markets. The Northern Ireland Department of Economic Development is currently developing proposals for the appropriate form of regulatory regime to facilitate the rapid development of the local gas market and promote competition in the wider energy scene.

STRUCTURE OF THE MARKET

10.6 The year 1994 saw a continuing extension of competition into the industrial market. The Government also outlined proposals to give domestic as well as industrial consumers a choice of suppliers.

10.7 When British Gas (BG) was privatised in 1986, it was given an effective statutory monopoly over supplies to premises taking less than 25,000 therms a year (equivalent to about 40 times an average household's consumption). Contract customers taking more than this were able, in theory, to buy their gas from other suppliers. Shortly afterwards, following complaints about anti-competitive practices by BG, the Monopolies and Mergers Commission (MMC) was asked to examine the contract market. In 1988 the MMC recommended, amongst other things, that BG should publish a schedule of contract gas prices and should not discriminate in pricing or supply; and that the company should contract initially for no more than 90% of any new gas field. They also concluded that further changes in the structure of the industry might be necessary if competition failed to develop over the subsequent first years of the new regime.

10.8 In 1991, the Office of Fair Trading (OFT) was asked to review progress towards a competitive market. It found that the steps taken in 1988 had been ineffective in encouraging self-sustaining competition, and recommended that BG should release gas to its competitors, agree to relax its tariff monopoly, and establish a separate subsidiary to operate its pipelines at arms' length from the rest of BG, with OFGAS regulation of its pipeline charges. Instead, BG undertook in March 1992 to allow competitors to take by 1995 at least 60% of the contract market above 25,000 therms (subsequently redefined as 45% of the market above 2,500 therms), and to release to competitors the gas necessary to achieve this; and to establish a separate transport and storage unit with regulated charges. At the same time, the Government took powers in the 1992 Competition and Service (Utilities) Act to reduce and/or remove the tariff monopoly; and in July 1992, it lowered the tariff threshold from 25,000 to 2,500 therms.

10.9 However, difficulties in implementing the March 1992 undertakings led to further references to the MMC in July 1992. After a year-long study, the MMC made new recommendations in 1993, including divestment by BG of its gas supply business, and a further reduction in the tariff threshold to 1,500 therms in 1997, with its removal some three to five years after divestment (ie by 2000-2002). The President of the Board of Trade decided in December 1993 to require full internal separation of BG's supply and transportation activities, but not divestment, and to accelerate removal of the tariff monopoly to March 1996, with a phased opening of the domestic market by the regulator over the following two years.

10.10 The introduction of effective competition into the contract market has taken both time and heavy regulatory intervention. By the end of 1994, competitors had indeed exceeded the target 45% of the market above 2,500 therms, but virtually all of this was in the firm gas market. Few inroads have yet been made into the interruptible market. Nevertheless, a start has been made in dismantling the regulatory levers in the contract market. In October 1994, OFGAS suspended for six months the requirement for BG to sell firm gas at schedule prices in the market above 25,000 therms, but retained the requirement for the smaller segment of the firm market (2,500-25,000 therms) and for interruptibles.

10.11 In March 1995, OFGAS set out criteria which BG must agree to meet if it wishes to stop pricing according to published schedules for the firm contract market between 2,500 and 25,000 therms and for interruptible supplies. Once these criteria are fully met, OFGAS will suspend for one year all the requirements on BG to publish schedules in the market above 2,500 therms, subject to no firm evidence emerging of predatory pricing. OFGAS will then conduct a further review to determine whether or not to remove the obligation permanently. OFGAS also decided, in agreement with the OFT, that BG's market share target of 55% should be formally removed, as it had already been reached; and that the suspension of price schedules in the over 25,000 therm firm gas market should be extended for 6 months, provided that BG continued to price in a non-discriminatory manner. In addition, the gas release programme will not be expanded in the year starting 1 October 1995.

THE GAS BILL

10.12 Legislation has been necessary to create the licensing framework for new entrants into the domestic market, and which would have the effect of separating BG's pipeline and supply businesses. In May 1994, DTI and OFGAS published a Joint Consultative Document setting out proposals for managing the introduction of competition into the domestic market, and in March 1995 the Government introduced a Gas Bill.

10.13 The Bill provides the legislative and regulatory framework for a restructuring of the gas industry so as to bring consumers the benefits of competition. In particular, it provides for the introduction of a new licensing framework, ie the separate licensing of "gas suppliers", whose function is to sell piped gas to consumers; "public gas transporters", whose function is to operate the pipeline system through which such gas will normally be delivered; and "gas shippers", whose responsibility is to arrange with public gas transporters for appropriate amounts of gas to be moved through the pipeline system.

10.14 The Bill makes provision for licences to be granted to these entities by the Director General of Gas Supply (DGGS) whose general duties are revised to take account of the new framework. It provides for such licences to be subject to standard conditions, except where circumstances dictate otherwise, and for these standard conditions to be capable of modification in certain circumstances.

10.15 In March 1995, the Government published details of the proposed licences, in order to give a clearer idea of the measures and safeguards that would be established for the operation of a fully competitive market in domestic gas supply. The principal Standard Conditions are likely to include:

- for supply licences, an obligation to supply all prospective domestic customers within the licensee's area (subject to certain technical exceptions); to publish charges; to offer a reasonable range of methods of payment to customers; to supply gas where directed to do so by the DGGS, to ensure the continuity of supply to customers in a "supplier of last resort" situation; to make financial arrangements, such as a bond, to cover the costs which would be involved if another company had to take over supplying their customers; to offer consumers reasonable rights to terminate contracts; to provide special services for disabled customers or pensioners free of charge; to adopt suitable methods for dealing with customers who have genuine difficulty paying their bills; to advise customers on energy efficiency; to check the background of officers authorised to enter customers' premises and to provide them with official identification;

- for transportation licences, requirements to keep separate accounts for transportation and storage businesses; to furnish the DGGS with details of charges, and both the DGGS and the Gas Consumers' Council with other information; to establish standards of performance and to provide compensation if they are not met; to establish a "network code" under which others may use the pipeline system; to make effective arrangements for the provision of emergency services to the public; to raise a levy, if necessary, to compensate any suppliers with disproportionate social obligations; and to obtain the DGGS's approval before disposing of key assets;

- for shipper licences, obligations as regards the use of pipeline systems; to assist the transporter during an emergency; to supply the DGGS with appropriate information.

At the same time, the Network Code, agreed between BG Transco and users of BG's pipeline network, is designed to ensure continuity of supply and access on non-discriminatory terms.

10.16 The Government intends to introduce competition in the domestic market on a phased basis between 1996 and 1998. The first phase, from April 1996, will comprise a pilot scheme covering some 500,000 homes in Cornwall, Devon, and Somerset. The scheme will be extended to 2 million homes in 1997, as a precursor to nationwide competition in 1998.

10.17 While competition in gas supply to final consumers is the primary objective, the Government also aims to encourage competition wherever possible throughout the gas supply chain. It welcomes the emergence of a largely unregulated short-term wholesale market: both the Government and OFGAS intend to help remove any obstacles to its further development. The Gas Bill also includes a duty upon the Regulator to promote competition in ancillary activities such as gas storage, meter-reading, and the provision of meters. Ministers have launched a separate initiative to simplify access by new gas shippers to offshore pipelines and the beach terminals into which they feed. The question of introducing competitor trading of capacity in parts of the onshore pipeline network remains under review.

GAS PRICES

10.18 Gas prices have fallen substantially in real terms since 1986: by over 39% in the contract market, and by $17\frac{1}{2}$% in the domestic market (after including VAT at 8%). Part of this is explained by the impact on North Sea gas prices of the 1986 fall in oil prices; but competition in the contract market and price regulation in the tariff market have made an increasing contribution.

10.19 Price reductions have been particularly marked in the firm gas market, where competitors have been offering prices 10% or more below the BG schedule price to capture market share. In the interruptible market, prices are driven partly by the price of heavy fuel oil (the normal alternative fuel at dual-fired sites) and partly by the need of suppliers to sell gas in the market outside the seasonal periods of peak domestic demand. In late 1993 and again in 1994, BG was able to increase its interruptible price schedules. By the end of 1994, BG still retained over 90% of the industrial interruptible market. However, the requirement upon independent suppliers from October 1995 to balance what they put into the pipeline network each day with what they take out, is likely to stimulate greater competitor interest in the interruptible market; and the emergence of a spot market giving clearer signals of the value of peak and non-peak gas should also help this process. Current HMIP guidelines limit the period each year for which gas supply to dual-fired premises may be interrupted: the Government is considering how far this may unduly constrain competition and innovation in the interruptible market.

10.20 In the tariff market, setting aside the introduction of VAT on domestic fuel, prices are continuing to fall as a result of the RPI-4 price cap (see paragraph 2.36): BG has begun to move towards more cost-reflective pricing within the overall price cap. In November 1994, it announced discounts of around 5% to customers paying by means of monthly direct debit, and has signalled its intention to extend discounts to other prompt payers in the near future. Additionally, the company has confirmed that it will not introduce charges for the special services which it is required (under the terms of its licence) to offer to elderly and disabled customers.

10.21 Prices at the beach rose in 1991, with increasing demand for gas for power generation (see Chart 10.2 overleaf). The price of gas from new UK fields has probably been above the price at which new gas is landed in mainland European markets. There are indications that UK prices are now weaker, but the integration of the UK and European markets via the interconnector (see paragraph 10.23) should serve to bring UK and European prices together. The prices paid by final consumers in Britain compare favourably with those paid by European consumers, as Table A24 in Appendix 1 shows. The latter are disadvantaged by the relative inefficiency and large profit margins of pipeline companies and distributors, particularly in those countries where there is little or no gas-to-gas competition.

EXTERNAL TRADE IN GAS

10.22 Since 1992, the UK has exported gas from the Markham field to the Netherlands, equivalent to about 1% of UK production. In 1994, exports increased by 40% as the Markham field production built up to plateau levels.

10.23 The year 1994 saw the completion of a gas pipeline between Scotland and the Republic of Ireland, which at present is being kept for standby use. Of greater strategic significance for the UK was the

Chart 10.2
British Gas weighted average cost of gas (WACOG)[1], 1980 to 1994

(1) Including the Government's levy on indigenous supplies.
(2) Calculated using the GDP deflator at market prices

Source: Department of Trade and Industry

decision in December 1994 by 9 major companies (Amerada Hess, British Gas, BP, Conoco, Elf, Distrigaz, Gazprom, National Power, and Ruhrgas) to build an interconnector to link the UK and European gas grids. The pipeline is expected to have a capacity of 20 bcm, and to be operational by late 1998. It will add market opportunities in Europe to those in the UK, and will accelerate the exploration and development of UKCS gas fields. This link to a larger market will tend to move UK beach prices into closer alignment with European beach prices, in terms of both level and movements. By providing the basis for cross-Channel swap deals and for increased flexibility in the spot market, it will generally strengthen and enhance the UK wholesale market. In the longer term, a grid-to-grid connector opens up the possibility of physical imports of gas, hence enhancing the UK's security of supply.

10.24 The UK also imports gas from Norway via a pipeline which was built in the 1970s to handle gas from the Frigg field. This pipeline, which has a capacity of around 10 bcm, is now virtually empty as the Frigg Field runs out. Imports during 1994 were some 32% lower than in 1993, accounting now for under 5% of UK demand. The UK has made clear that it is willing to facilitate use of that pipeline for other imports once the Frigg field runs out, provided that satisfactory arrangements for jurisdiction can be agreed.

CHAPTER 11

OIL AND GAS PRODUCTION

11 OIL AND GAS PRODUCTION

UK PRODUCTION AND RESERVES

11.1 Full details of oil and gas exploration and production over the last year, and estimates of reserves, may be found in the companion volume to this Report, *Oil and Gas Resources of the United Kingdom*. This chapter therefore provides a brief summary of the main features of the year.

11.2 A number of records were set in 1994: an all-time record for oil and gas production combined (20% up on 1993, and nearly a third up on 1992); record gas production for the fifth successive year (now 58% up on the 1989 figure); the highest annual oil production since 1986; the highest ever oil production in a quarter (Q4 1994); and the highest oil production in a month (December 1994). Chart 11.1 shows changes in production since 1970.

11.3 The main facts and figures were:

- a total of 127 million tonnes of oil was produced (27% up on 1993);

- there were 93 oilfields (73 offshore, 20 onshore) as at March 1995 (compared with 85 at the end of 1993);

Chart 11.1

Trends in the production of UKCS oil and gas, 1970 to 1994

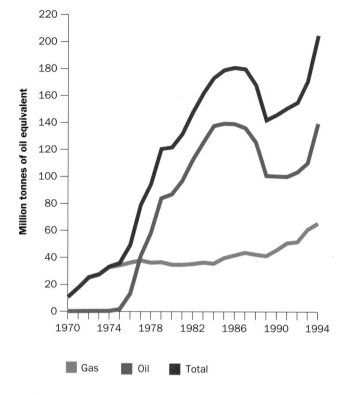

Source: Department of Trade and Industry

- gas production was over 70 billion cubic metres (7% up on 1993);

- there were 56 gasfields (53 offshore, 3 onshore) as at March 1995 (compared with 50 at the end of 1993);

- proceeds from the sale of oil and natural gas liquids (NGLs) produced from the UK Continental Shelf (UKCS) were estimated to be about £9.5 billion (up 10% on 1993);

- proceeds from the sale of gas produced from the UKCS were estimated to be about £3.8 billion (7% up on 1993);

- Government revenues from oil and gas were about £1.6 billion in 1994/95, compared with £1.3 billion in 1993/94;

- some 300,000 people are employed in oil- and gas-related activities, of whom about 27,000 work offshore (as at September 1994);

- twenty four new fields (9 oil, 13 gas, 2 condensate) were approved, the highest ever in one year.

11.4 For both oil and gas, production broadly matched the growth of reserves in 1994. The figures for reserves remaining (proven, probable, and possible) at the end of 1994 are 2,075 million tonnes for oil (compared with 2,100 million tonnes in 1993) and 1,910 billion cubic metres for gas (the same as in 1993). Recent trends are illustrated in Charts 11.2 and 11.3.

11.5 During 1994, 202 development wells began drilling offshore, a new record since drilling began on the UKCS in 1964. Onshore drilling activity was higher than in recent years, with 16 new wells. The most significant new development event of the year was the confirmation of a new hydrocarbon province west of Shetlands, with the approval of the Foinaven field. This is expected to be the first of a number of medium to large oil and gas fields in the region. First production from Foinaven is planned for 1996, using floating facilities, but it is hoped that

Chart 11.2

Discovered UK oil, cumulative production plus estimates of remaining reserves in present discoveries, 1980 to 1994

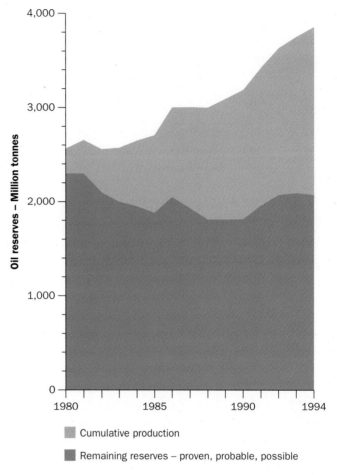

Source: Department of Trade and Industry

sufficient reserves will be proved in the next few years to justify pipeline evacuation. The largest single development approved in 1994 was the Britannia condensate field, with estimated reserves of 74 bcm of gas and 17 million tonnes of condensate.

11.6 Ninety nine exploration and appraisal wells were drilled in 1994, about 10% down overall on 1993, but within this total, exploration wells were more than 20% up, owing mainly to increased activity in the West of Shetlands and Southern Gas Basin areas. There were 13 "significant" discoveries (ie wells with a test flowrate of more than 1,000 barrels/day for oil, or 15,000 cubic feet a day for gas) announced in 1994, on acreage representing awards from eight licensing Rounds, including the first (1964) and the fifteenth (1994).

LICENSING

11.7 With the announcement of the 15th, 16th, and 17th Rounds of offshore oil and gas licensing during 1994, a new approach has been applied to licensing, with different parts of the UKCS offered in different Rounds to reflect the different stages of exploration reached. The 15th

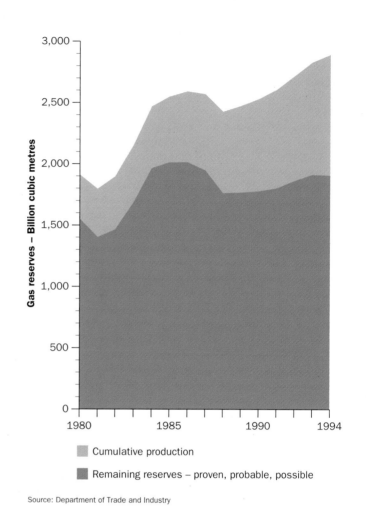

Chart 11.3

Discovered UK gas, cumulative production plus estimates of remaining reserves in present discoveries, 1980 to 1994

Source: Department of Trade and Industry

Round offered all available acreage in the main established areas of the Southern Basin and Central North Sea. This Round was announced in April, and the award of 29 blocks and part-blocks was announced in August. Blocks offered in the 16th Round (in November) covered areas around the coast of Great Britain and to the North of Scotland. These are areas where there has been some exploration in the past. Applications were received for 328 blocks, and 164 blocks were awarded. The 17th Round will cover frontier acreage where little exploration has taken place to date. Blocks for the 17th Round will be announced in the Summer of 1995.

11.8 New Onshore Licensing Regulations were made in April 1995. These create a unitary licensing regime, with a single licence (the Petroleum Exploration and Development Licence, or PEDL) replacing the former three licences covering the Exploration, Appraisal, and Development stages. The flexibility of a single licence covering all stages of activity will reduce the bureaucratic burden on licensees.

11.9 The new Onshore Regulations and revisions to the Offshore Regulations have been made to implement the EU Licensing Directive into UK law. The Directive extends the single market to the oil and gas sector by requiring that all licences should be offered on a non-discriminatory and transparent basis. The UK supported the Directive and welcomed its adoption. Implementation of this Directive also relieves UK oil and gas companies of some of the burdens under the EU Utilities Directive.

ENVIRONMENTAL MATTERS

UKOOA/JNCC Coastal Directory Regional Reports.

11.10 The 16th Licensing Round includes areas of environmental sensitivity. The DTI therefore felt that it would be advantageous if licence applicants had a standard baseline environmental document upon which they could base their work programmes.

11.11 DTI asked the Joint Nature Conservation Committee (JNCC) to prepare sixteen Coastal Regional Reports, with financial and project management assistance from the UK Offshore Operators Association (UKOOA).
Five regions were identified as being crucial to the 16th Round, and drafts of these five interim documents were prepared by mid-November 1994.
Final interim reports were made available to applicants and other interested bodies in January 1995.

11.12 Reports on the remaining 11 Regions will be produced over the next year, and all 16 reports will form part of a series of JNCC Coastal Directories.

CHAPTER
12

DOWNSTREAM OIL

12 Downstream Oil

DEVELOPMENTS IN UK MARKETS

12.1 Chart 12.1 shows the main oil products used in the UK. Transport fuels account for over two-thirds of energy use, with industrial fuel as a substantial, though rapidly shrinking, proportion. Lubricants are small in volume terms, but have a high added value. The production of chemical feedstocks using oil is another significant non-fuel market. Overall, the growth in the demand for oil products is relatively low, though there is substantial variation between products and sectors (see Chart 12.2). More information, including data since 1970, is given in Appendix 1 (Tables A14 and A15).

12.2 Petrol (gasoline) is the most important product in value and volume terms for the UK refining industry, accounting for about a third of total UK demand for oil products. Gasoline usage appears to have peaked, while demand for both automotive diesel and jet fuel are buoyant. The use of gas oil for heating and similar functions continues to fall gradually. Fuel oil demand is falling rapidly, principally because it is being displaced in the electricity generating market by gas. The use of orimulsion as a power station fuel has grown substantially since 1991, though demand in 1994 was below 1993 levels because one generating set that had previously used orimulsion was taken out of use at Richborough power station.

Chart 12.1
Inland use of petroleum products, 1994

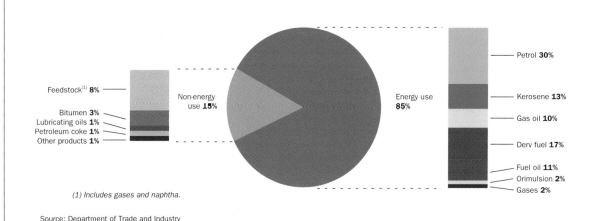

(1) Includes gases and naphtha.

Source: Department of Trade and Industry

TRANSPORT FUELS

12.3 Trends in the demand for transport fuels since 1990 are illustrated in Chart 12.2

Petrol

12.4 Petrol engines still dominate the UK car market. The transition to unleaded fuel is proceeding rapidly, and it now accounts for 60% of sales. Total petrol demand has been relatively stagnant in recent years, and in 1994 it declined by 4%, largely because of the continuing switch to diesel-engined cars and light vans (see below).

DERV

12.5 DERV demand continues to grow, and was about 700,000 tonnes higher in 1994 than 1993, with demand for use in cars and light commercial vehicles growing at around 8% a

Chart 12.2
Trends in demand for the main petroleum products, 1990 to 1994

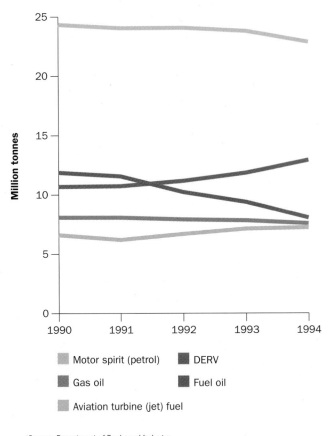

Source: Department of Trade and Industry

year. Chart 12.3 illustrates the growing demand for DERV since 1983. Total DERV consumption by heavy commercial vehicles has been restrained by the economic recession, but there has been a sharp increase in the proportion of new diesel cars and light commercial vehicles, and this change will become more marked as ageing petrol vehicles are scrapped. The increased popularity of diesel stems from a number of factors – improvements in the performance of diesel engines; their traditional economy; its cheapness relative to petrol (but see paragraph 12.19); and the public perception that they may be more environmentally-friendly. In 1991, only 3½% of the car stock was diesel engined; by 1994, the figure was 5% and growing, with diesels comprising about 20% of new car registrations. Diesel cars became popular in the rest of Europe many years ago, so that, despite the recent growth, DERV is still less important as a motor fuel in the UK than elsewhere, accounting for about 13% of product production by weight (1993) against the EC average of 19%.

Chart 12.3
Demand for DERV, 1983 to 1994

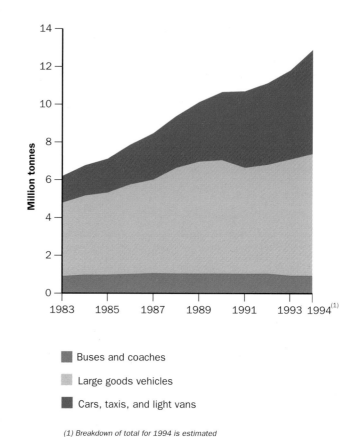

- Buses and coaches
- Large goods vehicles
- Cars, taxis, and light vans

(1) Breakdown of total for 1994 is estimated

Source: Department of Trade and Industry

Table 12.1
Super/Hypermarkets share of total UK deliveries

Year	Motor spirit	DERV
1990	8%	
1991	10%	1/2%
1992	12%	1%
1993	15%	2%
1994	18%	4%

Source: Department of Trade and Industry

Transport fuel retailing

12.6 Fuel retailing has seen some dramatic changes in the last decade. The number of filling stations has dropped by around 20%, from 22,000 in 1984 to 17,000 in 1994. Over the same period, the average throughput per station has increased by more than 40%, and is now amongst the highest in Europe. These changes have reduced delivery and operating costs per unit of sales.

12.7 Retail price competition is now fierce, following the entry of the supermarkets and hypermarkets into the retail petrol market. The big food retailers now have about 20% of the petrol market, and a small, but growing proportion of the DERV market (see Table 12.1) They have led the market with their aggressive pricing policies. The UK now has some of the cheapest motor fuel at the pump in Europe.

Jet fuel

12.8 The expansion of traffic at UK airports, particularly at Heathrow and Gatwick, has led to continued strong growth in demand for jet fuel. Stronger jet fuel demand will make it harder for the UK refining industry to meet the increasing demand for DERV, as the two products are made from the same part of the crude oil barrel.

OTHER FUELS

Gas oil

12.9 The use of gas oil has been declining slightly over the last few years, mainly because of increased competition from gas as a heating fuel.

Fuel oil

12.10 The fall in demand for high sulphur heavy fuel oil can partly be attributed to the increased use of gas for electricity generation and the consequent mothballing of oil-fired power stations. Oil burning for power generation was down 22% in 1994. The marine bunker market continues to be an outlet for high sulphur heavy fuel oil, but environmental restrictions and the availability of cheap alternatives point towards a continued long term decline in the demand for this fuel on land. Notwithstanding the long term trend, demand for high sulphur heavy fuel oil was strong in the winter of 1994/95, and the price differential between it and low sulphur fuel oil was eroded. Most low sulphur fuel oil produced in the UK is exported, mainly to Italy and Ireland. The latter market is likely to shrink, owing to increased competition from natural gas.

IMPLICATIONS FOR REFINERS

12.11 The increase in the demand for DERV and the stagnation in demand for petrol were not anticipated by the refiners, and there is now potentially a significant problem of mismatch between UK refining supply and inland demand. Past heavy investment in cracking capacity means that the UK's 11 principal refineries are not only amongst the most sophisticated in Europe but are also the most heavily geared to the production of petrol and other light products (32% of the barrel against a European average of 22%). It now appears unlikely that demand for petrol will grow. If consumption of DERV continues to rise rapidly in the UK, it will become increasingly difficult for UK refiners to satisfy demand.

12.12 The challenge posed by the growth in demand for aviation kerosene is similar to that for DERV, with the additional problem of supply capacity. There are pipeline delivery systems direct to the large airports, though this infrastructure may need to be expanded to cope with the continued growth of Heathrow and Gatwick.

PRODUCT EXPORTS

12.13 Product trade is becoming increasingly important, with substantial flows of products in and out of the UK. Net exports were worth about £1.2 billion in 1994, and have been following a generally upward trend. Exports were about a quarter of production in 1994, the principal destinations being France, Netherlands, Ireland, and Italy. Net exports in 1994 were slightly lower than in 1993 because unplanned refinery shutdowns led to substantial falls in production. The gross volume of trade undertaken is far greater than is suggested by the net figures in Table 12.2.

THE UK CRUDE SLATE

12.14 Changing patterns of demand have implications for the refiners' crude slate (the crude oils they refine). The light

Table 12.2
Petroleum product trade volumes, UK net exports

Million tonnes

	1990	1991	1992	1993	1994
Net exports[1] by product					
Motor spirit and aviation fuel	3.1	4.2	5.3	5.8	5.1
Gas oil/diesel oil	3.1	5.4	4.9	5.8	4.8
Fuel oil	-0.3	-0.3	-0.3	1.4	1.1
Other	1.1	1.3	1.4	1.8	3.4
Net exports[1] by country of destination/origin					
France	1.8	3.8	3.1	4.1	3.5
Germany	2.0	3.2	2.0	2.3	1.6
Irish Republic	2.6	2.5	3.0	3.1	3.3
Italy	2.0	2.0	2.6	3.7	2.7
Netherlands	-0.5	-0.2	1.1	1.8	2.4
Other European Union	-0.8	0.2	0.6	1.2	0.9
USA	1.2	0.9	0.6	0.9	2.1
Other[2]	-1.2	-1.9	-1.6	-2.2	-2.1
TOTAL	7.0	10.5	11.3	14.8	14.5

(1) Equals exports minus imports.
(2) Contains 1.4 million tonnes of imports for 1994 which cannot yet be allocated by country.
    A proportion will be imports from the named countries, thus reducing the net exports shown.
Source: Department of Trade and Industry.

North Sea crudes are ideal for petrol production, which currently accounts for about a third of the UK refiners' output. If demand were to continue to move towards heavier products, especially diesel, it might be more cost-effective to import more of the heavier Middle East and Russian crudes (which currently comprise 15% of the slate). This would mean the export of more North Sea crude, which is light with a low sulphur content, and commands a price premium over heavier oils. The net impact on the balance of payments would therefore be positive.

THE UK INDUSTRY

12.15 Independent research suggests that the UK refining industry is among the most efficient in Europe. UK refining capacity has undergone substantial rationalisation since the mid 1970s, with about 60 million tonnes/year of capacity being lost through the closure of three large and eight small refineries. As a consequence of these reductions, and of simultaneous growth in demand, capacity utilisation has increased dramatically, from about 60% in 1980 to 90% in 1994 (see Chart 12.4 overleaf). UK refineries are now middle-ranked in Europe in terms of age, using older equipment than those in Italy, Scandinavia, and Spain. Even so, they remain the most complex, following a doubling in secondary capacity (catalytic and/or hydrocracking) since 1980. This has resulted in by far the highest gasoline yield in the EC refining industry.

Price trends

12.16 Oil is one of the most international of industries, with both crude oil and products traded internationally.
In Europe, oil product prices are determined in the highly competitive Rotterdam cargo market.

12.17 In the world as a whole, product prices remain depressed because of a surfeit of refining capacity. Because prices are not fixed domestically, over-capacity in one country hits refiners elsewhere, reducing the profits of downstream oil companies.

12.18 In the last year, wholesale product prices have not shown a clear trend. Motor fuel prices rose in the first half of the year, before falling back in the autumn. Naphtha prices followed a generally upward trend for most of 1994, but have fallen back recently. The prices of Jet fuel/kerosene and gas oil have been more stable than those of most other fuels, but prices fell towards the year end. Fuel oil prices have been highly erratic over the year, with a marked closing in the differential between the two grades in the winter months.

12.19 Whilst wholesale prices are a significant aspect of the retail price, the tax element is extremely important, particularly in the case of motor fuels.

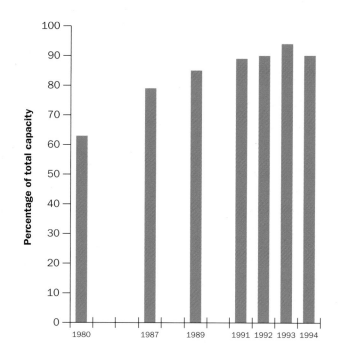

Chart 12.4

UK refinery capacity utilisation, 1980 to 1994

Source: Department of Trade and Industry

In the 1994 Budget, the duty differential between unleaded petrol and diesel (on a pence/litre basis) was closed, as the environmental case for giving diesel favourable treatment was no longer apparent. The Chancellor of the Exchequer's commitment to raise the duty on motor fuels by on average at least 5% each year in future Budgets means that the tax take per litre on motor fuels will continue on an upward path. But over the past few years, increases in duty have not led to an equivalent increase in pump prices, as pre-tax prices have fallen.

12.20 Table 12.3 suggests that the main oil products are generally cheaper in the UK than in most other major European countries, both including and excluding taxation and excise duties. The generally lower base prices (excluding tax and duties) reflect the efficiency of the UK refining and distribution sector, and the competitive nature of the market.

12.21 Despite its efficiency, the UK refining sector is currently barely covering variable costs. Recent estimates put margins at only $2/barrel in Europe, which barely covers variable costs. The tendency to world-wide over-capacity will weaken margins still further. Economic growth will increase UK demand for a wide range of oil products, boosting income, but investment to meet new environmental controls will increase costs.

Table 12.3
Pump price US$/gallon at mid December 1994

	Belgium	France	Germany	Italy	Holland	UK
Including tax & duty						
Four Star	4.77	4.68	4.82	4.72	5.24	4.13
Unleaded	4.29	4.42	4.42	4.40	4.80	3.74
Derv	3.42	3.20	3.19	3.46	3.37	3.76
Heating gas oil	1.02	1.67	1.17	3.40	1.61	1.00
Excluding tax & duty						
Four Star	1.21	0.88	1.07	1.13	1.22	1.02
Unleaded	1.20	0.94	1.01	1.16	1.20	1.03
DERV	1.19	0.92	0.98	1.03	1.13	1.04
Heating gas oil	0.77	1.00	0.79	0.98	1.00	0.77

Source: European Commission Oil Bulletin 15 February 1995

THE ENVIRONMENTAL CHALLENGE

12.22 The Eighteenth Report by the Royal Commission on Environmental Pollution "Transport and the Environment" set out the environmental problems caused by transport, many of which are related to transport fuels. These and other environmental concerns have led the EU to introduce stricter controls on pollution related to energy use. Meeting these new standards without losing competitiveness is the greatest challenge facing the refining industry.

12.23 More stringent environmental restrictions will affect the composition of oil products. The sulphur content of diesel is to be cut from 0.2% to 0.05% from October 1996. The production of leaded petrol is gradually being phased out, and steps are also being taken to reduce emissions of Volatile Organic Compounds (VOCs). At the same time, investment is needed to reduce pollution in the refining process, during product distribution, and at the point of sale. All these changes require the installation of new equipment, and may require new technical approaches.

12.24 Environmental regulations such as the Large Combustion Plant Directive have already had a significant effect at large industrial installations, including refineries and power stations. The Government is currently considering the most cost-effective means of achieving the further reductions in UK SO_2 emissions, as required by the UK's commitment to the new UNECE SO_2 protocol (see paragraph 6.31).

12.25 Investments for environmental control measures do not contribute directly to profit, and consequently depress the downstream oil industry's profitability. There is a view within the industry that the EU is placing itself at a competitive disadvantage, particularly in relation to the Far East, through agreeing to meet tighter environmental standards. The UK Government's approach is to ensure that proposals for more stringent environmental controls are based on a sound scientific analysis of the environmental impact, and are also subject to rigorous cost/benefit analysis.

Annex

Summary of Main Events in 1995 and 1996

COMPETITION

1995: passage of Gas Bill through Parliament.

October 1995: implementation of British Gas Network Code.

April 1996: pilot scheme in Cornwall, Devon, and Somerset for competition in domestic gas market (500,000) consumers.

ENVIRONMENT

March/April 1995: first Conference of Parties to UN Framework Convention on Climate Change.

1995: second report by Intergovernmental Panel on Climate Change (IPCC).

July 1996: second Conference of Parties to Convention on Climate Change.

1996: first draft of second UNECE NOx protocol.

UK ENERGY POLICY

March 1995: 16th offshore Licensing Round completed.

Summer 1995: blocks announced for 17th round.

1995: publication of offshore abandonment guidelines.

1995: publication of Nuclear Review.

1995: sale of British Coal's remaining non-mining assets.

1996: fourth NFFO Order for England and Wales, second NFFO order for Northern Ireland, second Scottish Renewables Order.

EU

July 1995: European Commission to review Large Combustion Plant Directive.

1996: Intergovernmental Conference (IGC): possible EU Energy White Paper/proposals for Energy Chapter to Treaty on European Union.

ENERGY EFFICIENCY

Increased budget for Home Energy Efficiency Scheme (HEES) to £100 million a year from 1 April 1995.

New Small Company Environmental and Energy Management Scheme (SCEEMAS) from May 1995.

OTHER

1995: UK to ratify Nuclear Safety Convention.

1995: negotiations on second Energy Charter Treaty on national treatment for the admission of investments.

1995: Privatisation of AEA Technology, the commercial activities of the UK Atomic Energy Authority.

Appendices

APPENDIX 1   PRODUCTION AND CONSUMPTION OF ENERGY IN THE UNITED KINGDOM

A1.1 This Appendix provides a statistical overview of energy supply and demand in the UK. It examines the latest figures in production and consumption, in foreign trade in fuels and in fuel prices. Details are also presented of energy balances in other countries. Further information concerning individual fuels is given in the chapters of this Report dealing with the energy industries.

PRIMARY ENERGY PRODUCTION AND CONSUMPTION

A1.2 Energy sources can be considered as primary or secondary. Primary fuels either occur naturally, as in the case of crude oil, natural gas or coal, or are derived by directly harnessing naturally occurring energy, as with nuclear or hydro electricity. Secondary fuels, such as petroleum products, coke and

Chart A1.1

UK energy flows, 1994 (million tonnes of oil equivalent)

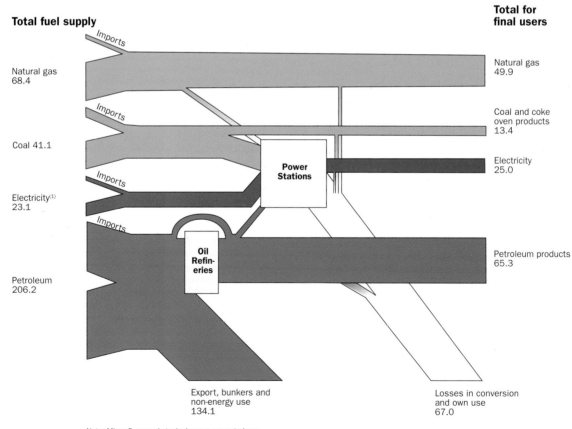

Note: Minor flows and stock changes are not shown
(1) Primary electricity includes nuclear, hydro, wind stations and imports.

Source: Department of Trade and Industry

secondary electricity, are obtained either from primary fuels or from other secondary fuels, by conversion processes.

A1.3 The flow of energy in the UK from primary production and imports, through the conversion industries and to final users is illustrated in a simplified form in Chart A1.1. It can be seen that final energy demand is met by a combination of primary and secondary fuels.

A1.4 As can be seen in Chart A1.1, the conversion of fuels, such as coal, oil and gas, to electricity leads to substantial losses of energy. In 1994 49 million tonnes of these three fuels (in oil equivalent terms - mtoe) were used by the major power producers to produce $17\frac{1}{2}$ mtoe of electricity, a thermal efficiency of 36 per cent. Final users also received a further $8\frac{1}{2}$ mtoe of electricity from nuclear, hydro and wind sources and from imports. This means that in 1994, for every toe of electricity generated by major power producers, 1.9 toe of coal, oil or gas were consumed at the power stations. If a final user switches from coal, oil or gas to electricity there will usually be an increase in the use of these fuels by power stations, unless additional electricity comes from nuclear,

Chart A1.2

UK production of primary fuels, 1970 to 1994

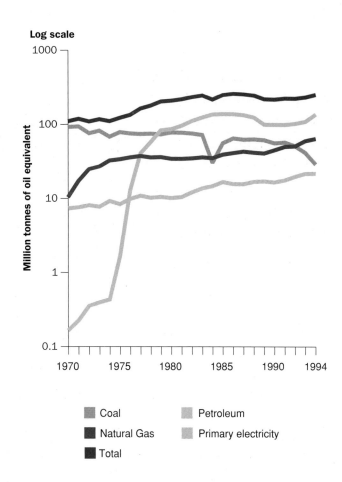

Source: Department of Trade and Industry

Chart A1.3
Primary energy supply by fuel, 1994

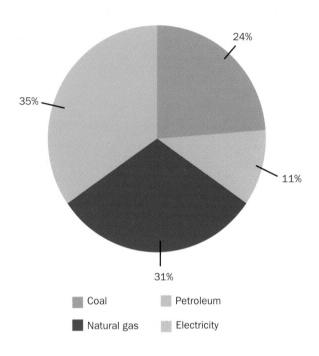

Source: Department of Trade and Industry

hydro, wind or imported sources. This needs to be borne in mind when analysing trends in final energy consumption.

A1.5 The UK is fortunate in possessing a range of different primary fuels. Chart A1.2 shows trends in the production of these fuels and illustrates the rapid increase in the production of North Sea oil and natural gas in the 1970s. Figures for the production of primary fuels are shown in Table A1. Chart A1.3 shows the contribution of these fuels to primary energy supplies in 1994. Supplies are met from domestic production and from imports. Figures for earlier years are presented in Table A2.

A1.6 Unlike most of its competitors the UK is self-sufficient in energy. Chart A1.4 shows the UK's primary energy production and consumption, and illustrates the degree to which the country was dependent on energy imports prior to North Sea oil and gas becoming available.

APPENDIX 1

Chart A1.4
UK primary energy production and consumption, 1970 to 1994

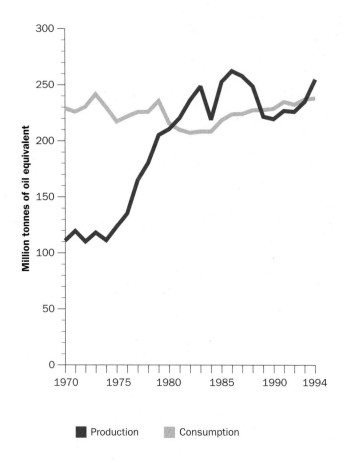

Source: Department of Trade and Industry

In the early 1970s energy imports accounted for over 50% of the UK's consumption, but in 1983 the UK was a net exporter, at a level equivalent to 18% of inland consumption. After 1983, net exports declined slowly, and, following temporary production losses in the North Sea since 1988, the UK was a net importer of energy until 1993. The balance switched again in 1994 as production on the UK continental shelf recovered.

A1.7 Chart A1.5 shows the balance between production and demand separately for oil, gas and coal for selected years between 1970 and 1994. By 1994 the UK was virtually self-sufficient in gas and more than self-sufficient in oil, but dependent on coal imports.

Chart A1.5

Oil, gas and coal production and primary demand, 1970, 1982 and 1994

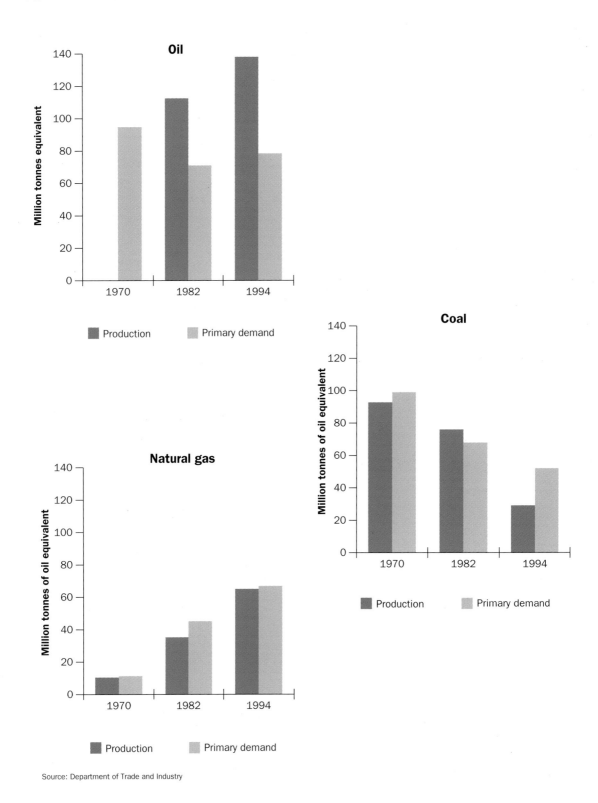

Source: Department of Trade and Industry

APPENDIX 1

IMPORTS AND EXPORTS

A1.8 The domination of the UK's foreign trade in fuels in 1994 by crude oil and petroleum products is illustrated in Chart A1.6. Figures for earlier years are given in Table A3. In 1994 the UK had a surplus in volume terms equivalent to 40 million tonnes of oil. When expressed in financial terms this amounts to a surplus of £3.4 billion in 1994, more than £2 billion higher than in 1993. Figures for the value of trade in recent years are presented in Table A1.1.

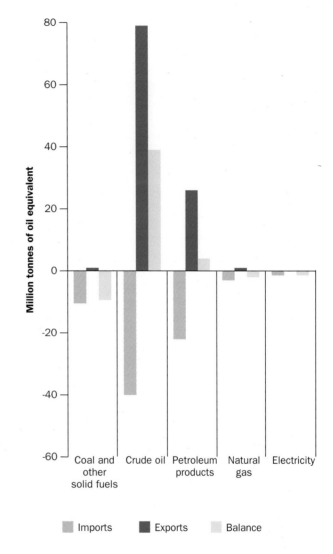

Chart A1.6
Imports and exports by fuel, 1994

Source: Department of Trade and Industry

Table A1.1
Value of UK imports and exports of fuels, 1988 to 1994[1]

£ million

	Imports	Exports	Net exports
1988	4,675	6,257	1,582
1989	6,071	6,172	101
1990	7,418	7,771	353
1991	7,165	7,107	−58
1992	6,620	6,879	259
1993p	6,959	8,217	1,258
1994p	5,720	9,080	3,360

(1) All values on a 'free on board' basis, i.e. with imports adjusted to exclude the costs of insurance, freight etc.
Source: Department of Trade and Industry

FINAL ENERGY CONSUMPTION

A1.9 Table A4 shows how demand for energy by final users has been met by the various primary and secondary fuels since 1970. The split of final energy consumption between the demands of households, industry, transport and other sectors (government, commerce and agriculture) is given in Table A5 and is illustrated in Chart A1.7 for 1994. The transport sector is responsible for a third of all final energy consumption.

A1.10 The level of demand depends upon the price of energy, the uses to which it is to be put, and the technical efficiency of the processes involved in its use. The way demand is split between fuels depends on the availability of supplies, the relative cost of different fuels, and technical needs. Significant shifts in the pattern of demand and supply have developed in response to altered needs, new supplies and social and environmental pressures.

Domestic consumption

A1.11 Since the mid 1970s gas has replaced coal as the main household fuel; this can be seen in Chart A1.8 and Table A6.

Chart A1.7
Final energy consumption by sector, 1994

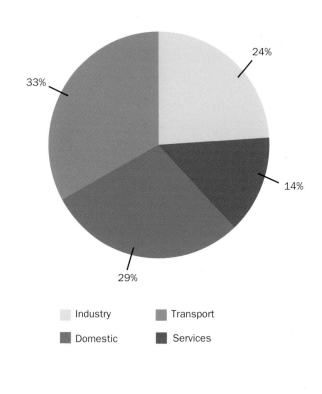

Source: Department of Trade and Industry

In contrast consumption of electricity and petroleum in the home have changed only gradually over this period.

A1.12 More than half the energy used by the domestic sector is for space heating, with a further quarter used to heat water (Chart A1.9). This pattern has changed little since 1970, although the energy used for lights and appliances has doubled, whilst the energy used for cooking has fallen gradually. Natural gas is by far the

APPENDIX 1

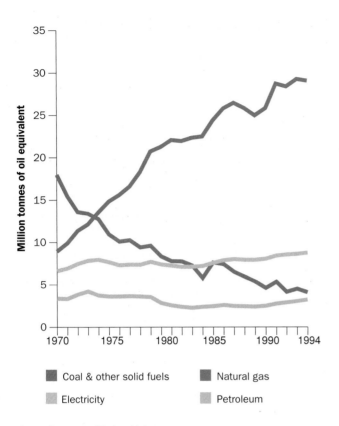

Chart A1.8
Domestic energy consumption by fuel, 1970 to 1994

Source: Department of Trade and Industry

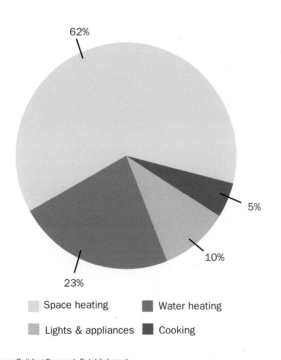

Chart A1.9
Domestic energy consumption by end use, 1991

Source: Building Research Establishment

143

dominant heating fuel with 74% of the market; electricity accounts for 12%, with solid fuels and oil accounting for 9% and 3% respectively.

A1.13 By 1991 more than 80% of homes had central heating. Chart A1.10 illustrates trends in the ownership of central heating and consumer durables since 1970. It can be seen that the household penetration of refrigerators and washing machines, the most energy-demanding domestic appliances, is levelling off. The ownership of tumble-dryers, dishwashers and microwave ovens is still rising rapidly.

A1.14 Some 5% of average household expenditure is on fuel, light and power. However, this average figure conceals considerable variation between households;

Chart A1.10
Trends in ownership of consumer durables, 1970 to 1993

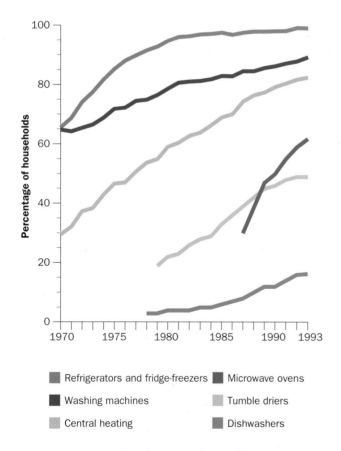

Source: Central heating, washing machines, refrigerators, fridge-freezers (Family Expenditure Survey), tumble dryers, dishwashers, microwave ovens (up to 1992 General Household Survey, 1993 Family Expenditure Survey).

APPENDIX 1

Table A1.2
Household expenditure on fuel, light and power by gross household income in the UK, 1993

Gross income decile group	Average weekly expenditure (£)		Expenditure on fuel, light and power as percentage of total
	Total	On fuel, light and power	
Lowest ten per cent	80.35	9.30	11.6
Second decile group	116.86	11.48	9.8
Third decile group	154.21	11.93	7.7
Fourth decile group	192.88	12.21	6.3
Fifth decile group	238.22	12.66	5.3
Sixth decile group	265.43	13.13	4.9
Seventh decile group	306.51	13.42	4.4
Eighth decile group	350.11	14.40	4.1
Ninth decile group	447.96	15.40	3.4
Highest ten per cent	614.21	18.49	3.0
All households	**276.68**	**13.24**	**4.8**

Source: Family Expenditure Survey

lower income households tend to spend a much higher proportion of their income on fuel than wealthier households. This is illustrated in Table A1.2, which summarises the findings of the 1993 Family Expenditure Survey.

Industrial consumption

A1.15 The demands of industry for energy are very diverse. Table A17 shows energy costs as a proportion of total costs for the major industrial groups. The most energy intensive industries include the water industry and the sectors making paper, bricks, metals, chemicals and cement. The sectors for which energy costs are least important include the clothing, printing and vehicle industries.

A1.16 Industry uses energy as a source of power, heat, and light. In addition, oil and gas are feedstocks for the manufacture of chemicals, and coke is central to the manufacture of steel. Table A8 shows the uses to which energy is put in the various sectors of industry. It can be seen that industries, such as the iron and steel, glass, cement and brick making sectors, use energy mostly for high temperature processes. Many of the other sectors use energy for low temperature processes, space heating, drying or motors.

A1.17 Structural and technological change has had a marked impact on industrial energy consumption. Overall consumption by industry has fallen by 40% since 1970, with the demand for coal falling by 71% and the demand for petroleum by 70% over this period. In contrast the demand for electricity rose by 33% (see Chart A1.11 and Table A7). Natural gas became the dominant fuel in the early 1980s and has retained this role since then.

A1.18 In addition to the energy purchased and consumed directly by industry, each manufacturing industry also depends upon the energy required to produce its non-energy inputs. This indirect energy consumption can be estimated by energy input-output analysis, allowing those industries, for which indirect consumption is important, to be separated from those which consume most of their total energy needs directly. For example, in the manufacture of cement, earthenware, paper and glass more than a third of the total energy needs are consumed directly by these industries. In contrast, in the production of industrial and office machinery, motor vehicles, electrical goods and furniture around 90 per cent of the total energy needs are consumed indirectly during the manufacture of the required input materials.

Chart A1.11
Industrial energy consumption by fuel, 1970 to 1994

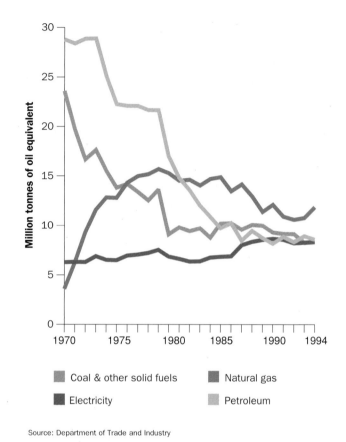

Source: Department of Trade and Industry

Transport

A1.19 Energy consumption by the transport sector increased by 79% between 1970 and 1994, and now accounts for a third of all final consumption (Chart A1.7). Demand is dominated by road transport fuels, particularly petrol and diesel, although electricity is used by railways. The growth in transport demand has been continuous,

APPENDIX 1

reflecting rising incomes and the changing patterns of domestic and commercial activity.

A1.20 Industrial demand for transport fuels has also altered, with the growth of road freight reflecting changing retail distribution systems and the decline of those industries which had previously made the greatest use of the railways.

Commercial and public buildings

A1.21 As the service sector (which covers the commercial sector and public administration) has grown it has naturally increased its share of total energy demand. Service industries do not have the very heavy energy demands of some manufacturing industries, but energy used for heating, lighting, air conditioning, and running computers is a significant part of their costs.

A1.22 Like the domestic sector more than half the energy used in commercial and public buildings is for space heating (see Chart A1.12). Table A9 provides a breakdown of the estimated energy consumption in buildings used for different purposes in 1991. Overall, the service sector is less dependent on gas than the domestic sector; offices and shops tend to use more electricity than gas.

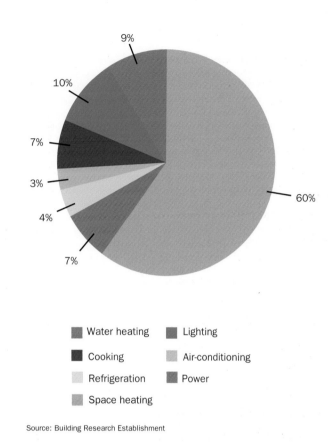

Chart A1.12
Energy consumption in the service sector, 1991

- Water heating
- Cooking
- Refrigeration
- Space heating
- Lighting
- Air-conditioning
- Power

Source: Building Research Establishment

PRICES

A1.23 Energy prices are set by the interaction of demand and supply, though not always in fully competitive markets. Where fuels are readily transported, it is world demand and supply which matters. Two primary fuels - oil and coal - are traded on world markets, and the world prices of these fuels influence, but by no means determine, the prices which can be charged for fuels that can be sold only on national or regional markets - notably gas. Trends in crude oil prices are illustrated in Chart A1.13. Trends in the spot price of coal are shown in Chart 8.4 in Chapter 8.

A1.24 The final prices paid by customers reflect many influences other than international energy prices, most obviously the costs of electricity generation or oil refining, and the costs of distribution, retailing and marketing. They also include any taxes and levies. Overall trends in domestic and

Chart A1.13
Trends in the price of crude oil, 1972 to 1994

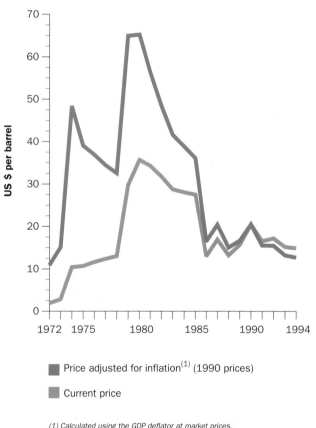

■ Price adjusted for inflation[1] (1990 prices)
■ Current price

(1) Calculated using the GDP deflator at market prices.

Source: BP Statistical Review of World Energy, 1994

APPENDIX 1

industrial prices are considered below. Trends in the prices of individual fuels are discussed in more detail in the chapters of this Report dealing with the energy industries.

Domestic prices in the UK

A1.25 Chart A1.14 shows the course of domestic fuel prices over the period 1980 to 1994, whilst some typical retail prices are presented in Table A18. Domestic electricity prices rose by about 6% more than inflation (as measured by the GDP deflator) between 1989 and 1992. This trend was reversed in 1993, with a ½% fall in nominal prices and a 4% fall in real prices. During 1994 prices started to rise again as a result of the introduction of 8% VAT on domestic fuel in April, although prices excluding VAT fell by 4 per cent in real terms.

Chart A1.14
Domestic fuel prices indices in real terms[1], **1980 to 1994**

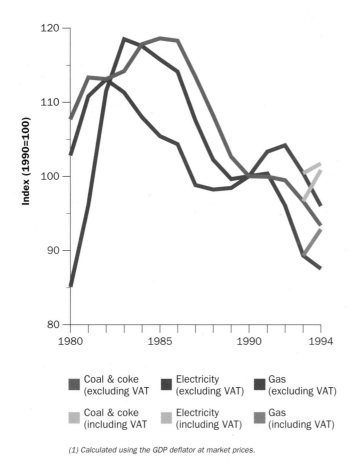

(1) Calculated using the GDP deflator at market prices.

Source: Department of Trade and Industry

A1.26 Chart A1.15 shows the movement in the price of oil and petrol supplied to the domestic sector since 1980: prices fell sharply between 1985 and 1988. Petrol prices are, of course, affected by more than the movement in crude oil prices. There have been changes in retailers' margins, but petrol duty and VAT now constitute nearly 75% of the price of unleaded petrol. Retail prices of petroleum products since 1970 are given in Table A19 and details of the rates of duty charged on the different fuels are in Table A20.

Industrial prices in the UK

A1.27 Chart A1.16 shows the movement of industrial fuel prices over the period since 1980. The overall index of fuel prices has fallen steadily in real terms since 1985: by 1994 prices were less than three-fifths of their 1985 level. Table A1.3 shows the history for the past 7 years in real terms. Table A21 gives prices by size of consumer since 1970.

Chart A1.15

Domestic fuel price indices in real terms[1], 1980 to 1994

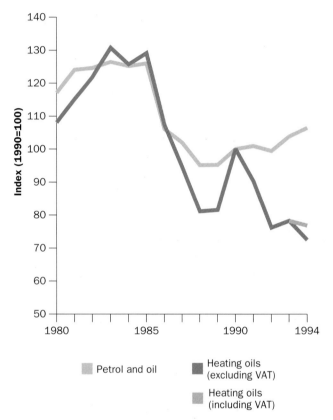

(1) Calculated using the GDP deflator at market prices.

Source: Department of Trade and Industry

APPENDIX 1

Chart A1.16
Industrial fuel price indices in real terms[1], 1980 to 1994

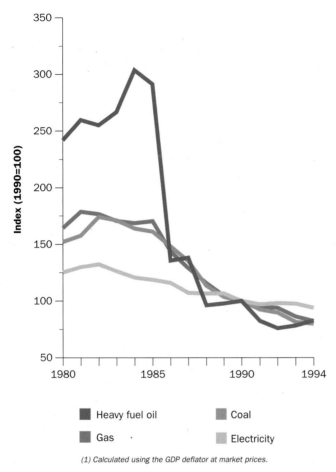

(1) Calculated using the GDP deflator at market prices.

Source: Department of Trade and Industry

Table A1.3
Fuel prices paid by manufacturing industry in the UK in real terms[1]

Index (1990 = 100)

	Coal	Heavy fuel oil	Gas	Electricity	Total fuels
1988	113	96	116	106	107
1989	103	98	105	107	105
1990	100	100	100	100	100
1991	92	82	95	97	94
1992	90	76	94	98	94
1993	82	78	86	98	92
1994	79	83	81	94	89

(1) Calculated using the GDP deflator at market prices.
Source: Department of Trade and Industry

INTERNATIONAL COMPARISONS

Methodological changes in the UK

A1.28 In order to bring the United Kingdom's energy statistics more into line with those published by the international energy organisations the Department revised its statistical methodologies in July 1994. The two main changes introduced are described below. A more detailed explanation can be found in the 1994 edition of the Digest of United Kingdom Energy Statistics.

A1.29 The changes were:

- to adopt the tonne of oil equivalent (toe) as the common unit of measurement, replacing the therm. The internationally accepted definition (1toe = 10^7 kilocalories) is now used.

- to move from a substitution basis to an energy supplied basis for expressing the contribution of nuclear and hydro electricity, onshore wind and electricity imports.

A1.30 A third proposed change, the move from gross to net calorific values, was considered in 1994 but not adopted. The Department has therefore continued to use gross calorific values to convert data on individual fuels to a common unit of energy, whereas the Statistical Office of the European Communities (Eurostat) and the International Energy Agency (IEA) use net calorific values. (Net values exclude the amount of heat needed to evaporate the water present in the fuel or formed during combustion.) This remains the main difference between the overall energy statistics published by the Department and those published by Eurostat and the IEA.

Energy balances compiled by Eurostat

A1.31 Eurostat compiles monthly and annual energy statistics for Member States of the European Union. A summary of the Eurostat's latest complete figures are presented in Table A22. It can be seen that the UK is far less dependent on imports than are our EU partners.

A1.32 Further details and technical notes can be found in Eurostat's energy publications:

- Energy – Monthly Statistics
- Energy – Yearly Statistics
- Energy – Energy Balance Sheets
- Energy in Europe: Annual Energy Review
- Energy in Europe: Energy Trends and Policies in the European Union (annual)

Energy balances compiled by the IEA

A1.33 The IEA compiles balances both for OECD countries and non-OECD countries. These are published annually (see below). Data for a selection of OECD countries are summarised in Table A23. Norway, Australia and Canada had large trade surpluses in fuels in 1992.

A1.34 Relevant IEA publications are:

- Energy Balances of OECD Countries (annual)
- Energy Statistics of OECD Countries (annual)
- Energy Statistics and Balances of Non-OECD Countries (annual)
- Energy Policies of IEA Countries (annual)

International Energy Prices

A1.35 Comparisons of prices in different countries can be difficult: International price comparisons are affected by the exchange rates ruling at the time the comparison is made. There are also differences in taxation between countries which can affect prices paid by purchasers. There are two main sources of data available for comparing prices across different countries. Each attempts to overcome the problems of comparing international prices. These two sources are the IEA's quarterly prices and the EC price transparency directives.

A1.36 The IEA compiles and publishes quarterly tables of average prices paid by industry and households for fuels. Prices of gas and electricity for industrial and domestic use for EU and selected OECD countries are summarised in Table A.24. Compared with the OECD countries listed, the UK is in the middle of the range of average prices. Compared with other EU members in 1993 the UK had the second lowest gas prices and household electricity prices, and was in the middle of the range of average industrial electricity prices.

A1.37 Generally United Kingdom industrial electricity prices at the beginning of July 1994 (including all taxes and duties except VAT) were amongst the lowest in Europe. For example, for consumers of 30,000 kWh a year, 9 of the other 11 member states had higher electricity prices than the UK; for consumers of 2 GWh a year, 8 of the other 11 member states had higher electricity prices than the UK; and for consumers of 50 GWh a year, 9 of the other 11 member states had higher electricity prices than the UK. A similar but less consistent pattern emerges for industrial gas prices. For example, United Kingdom gas prices at the beginning of July 1994 were the lowest in Europe both for the smallest customers (116,300 kWh a year), and for middle ranking customers (1.163 GWh a year). For larger customers of 11.63 GWh a year only 7 of the other 11 member states had higher gas prices than the UK, while for very large customers (116.3 GWh a year) the UK had the third highest prices out of the 9 member states recording gas customers of that size.

A1.38 The following publications contain further details:

- IEA Statistics, Energy Prices and Taxes, Third Quarter 1994

- Eurostat - Electricity Prices 1990-1994, Gas Prices 1990-1994 - as updated by Rapid Reports: Energy and Industry

TECHNICAL NOTES

A1.39 This section explains the technical basis of the overall energy statistics presented in this Appendix and elsewhere in this Report. More detailed statistics and further technical details can be found in the 1994 *Digest of UK Energy Statistics*.

Units of measurement

A1.40 A common unit of measurement is used to enable fuels to be compared and aggregated. The unit used by the Department is the tonne of oil equivalent. For consistency with the IEA and Eurostat this is defined as follows:

APPENDIX 1

1 tonne of oil equivalent
- $= 10^7$ kilocalories
- $= 397$ therms
- $= 41.868$ GJ
- $= 11{,}630$ kWh

Other conversion factors are given at the end of these notes.

A1.41 The tonne of oil equivalent (toe) is a measure of energy content rather than a physical quantity. Data on individual fuels are converted from their original units to toe using the appropriate gross calorific values for each fuel and the conversion factors shown above. The methods used to obtain figures on this 'heat supplied' basis are explained more fully in the 1994 *Digest of UK Energy Statistics*.

A1.42 Primary fuel inputs for nuclear, hydro and net imports of electricity are expressed in oil equivalent terms on an energy supplied basis, that is in terms of the energy content of the electricity produced.

Primary consumption (Table A2)

A1.43 Energy consumption is usually measured in one of two different ways. The first, known as the primary fuel input basis, assesses the total input of primary fuels and their equivalents into the economy. This measure therefore includes energy used and lost in conversion of primary fuels to secondary fuels (for example in power stations and refineries), energy lost in the distribution of fuels (for example in transmission lines) and energy conversion losses by final users.

Final consumption (Tables A4 to A7)

A1.44 The second method of measuring energy consumption is known as consumption by final users. This approach measures the inputs of primary and secondary fuels to final users. The figures are therefore net of the fuel industry's own use and conversion, transmission and distribution losses but include conversion losses by final users.

A1.45 Figures for final energy consumption relate to deliveries if fuels can be stored by the users and if actual consumption data are not available. Final consumption of petroleum and solid fuels is on a deliveries basis throughout except for the use of solid fuels by the iron and steel industry. Figures for the domestic use of coal are based on deliveries to merchants.

A1.46 Figures for final consumption of electricity include sales through the public distribution system and, from 1990, consumption by generators other than the major power producers. Thus electricity consumption from 1990 includes that produced by industry; figures for deliveries of other fuels to industry therefore exclude amounts used to generate electricity from 1990.

Symbols used

A1.47 The following symbols are used in this appendix:

.. not available
– nil or less than half the final digit shown
p provisional

STANDARD CONVERSION FACTORS

1 tonne of oil equivalent	= 10^7 kilocalories	1 kilowatt (kW)	= 1,000 watts
	= 397 therms	1 megawatt (MW)	= 1,000 kilowatts
	= 41.868 gigajoule	1 gigawatt (GW)	= 1,000 megawatts
	= 11,630 kilowatt hours	1 terawatt (TW)	= 1,000 gigawatts
1 therm	= 29.3071 kilowatt hours	1 petawatt (PW)	= 1,000 terawatts
1 gigajoule	= 9.4781 therms		
1 tonne of UK crude oil	= 7.55 barrels	1 gallon (UK)	= 4.54609 litres

APPENDIX 1

Table A1
Production of primary fuels and equivalents (heat supplied basis), 1970 to 1994 – United Kingdom

Thousand tonnes of oil equivalent

	Coal	Petroleum	Natural gas	Nuclear electricity	Natural flow hydro electricity	Total
1970	92,792	166	10,461	6,998	390	**110,807**
1975	79,172	1,675	34,203	8,120	326	**123,496**
1980	78,502	86,911	34,790	9,909	338	**210,450**
1985	56,572	139,404	39,679	16,499	351	**252,506**
1990	56,641	100,332	45,566	16,257	454	**219,256**
1991	57,963	100,085	50,733	17,431	403	**226,621**
1992	51,803	103,435	51,597	18,454	475	**225,770**
1993	42,056	110,105	60,682	21,492	393	**234,736**
1994p	29,821	138,431	65,399	21,184	445	**255,280**

Table A2
Inland consumption of primary fuels and equivalents (heat supplied basis), 1970 to 1994 – United Kingdom

Thousand tonnes of oil equivalent

	Coal	Petroleum	Natural gas	Nuclear electricity	Natural flow hydro electricity	Net imports of electricity	Total for energy use	Gross inland consumption[1]
1970	98,994	94,752	11,300	6,998	390	48	**212,482**	229,062
1975	73,716	86,298	35,047	8,120	326	8	**203,514**	217,341
1980	73,263	77,358	44,785	9,909	338	–	**205,654**	215,680
1985	64,824	73,477	51,803	16,499	351	–	**206,955**	218,423
1990	67,650	77,976	50,445	16,257	454	1,027	**213,825**	228,579
1991	67,126	78,114	53,959	17,431	403	1,411	**218,450**	234,685
1992	63,449	77,802	54,766	18,454	475	1,435	**216,387**	232,324
1993	55,129	79,681	62,386	21,492	393	1,437	**220,527**	236,862
1994p	51,816	76,828	67,686	21,184	445	1,452	**219,410**	235,974

(1) Includes non-energy uses and marine bunkers

Table A3
Volume of imports and exports of fuels and related materials[(1)], 1970 to 1994 – United Kingdom

Million tonnes of oil equivalent

	Coal and other solid fuel	Crude oil	Petroleum products	Natural gas	Electricity	Total
Imports						
1970	0.1	107.9	25.0	0.9	0.1	134.0
1975	3.5	93.3	17.6	0.9	–	115.3
1980	4.7	48.0	15.7	9.7	–	78.1
1985	8.3	28.8	27.2	12.6	–	76.9
1990	10.2	47.8	25.2	7.3	1.1	91.6
1991	13.5	50.1	24.0	6.5	1.4	95.5
1992	14.2	51.3	22.3	5.5	1.4	94.7
1993p	13.0	53.3	21.3	4.3	1.4	93.2
1994p	11.3	44.2	20.3	3.0	1.5	80.1
Exports						
1970	2.9	1.2	18.7	–	–	22.8
1975	2.4	0.9	15.5	–	–	18.8
1980	3.3	41.2	17.7	–	–	62.2
1985	2.4	85.2	21.0	–	–	108.6
1990	1.9	59.2	22.5	–	0.1	83.6
1991	1.5	56.6	25.0	–	–	83.1
1992	0.8	58.6	26.1	–	–	85.5
1993p	1.0	66.6	29.3	0.5	–	97.5
1994p	1.2	87.7	29.9	1.0	–	119.8
Net exports						
1970	2.8	106.7	–6.3	–0.9	–0.1	–111.2
1975	–1.1	–92.4	–2.1	–0.9	–	–96.5
1980	–1.4	–6.8	2.0	–9.7	–	–15.9
1985	–5.9	56.4	–6.2	–12.6	–	31.7
1990	–8.3	11.4	–2.7	–7.3	–1.0	–7.9
1991	–12.0	6.5	1.0	–6.5	–1.4	–12.4
1992	–13.4	7.3	3.8	–5.5	–1.4	–9.2
1993p	–12.0	13.3	8.1	–3.7	–1.4	4.2
1994p	–10.1	43.6	9.6	–2.0	–1.5	39.6

(1) The figures generally correspond to those published in section 3 of the Overseas Trade Statistics of the United Kingdom.

APPENDIX 1

Table A4
Final consumption of energy by fuel (heat supplied basis), 1970 to 1994 – United Kingdom

Thousand tonnes of oil equivalent

	Coal	Other solid fuels[1]	Coke oven gas	Natural gas[2]	Electricity	Petroleum	Total[3]
1970	29,822	15,134	1,164	14,769	16,541	68,512	**146,344**
1975	16,172	9,368	1,038	31,674	18,293	64,802	**141,417**
1980	12,854	5,479	642	42,015	19,278	62,408	**142,711**
1985	12,124	6,446	768	46,795	20,119	56,416	**142,684**
1990	8,115	6,106	602	45,419	23,593	63,235	**147,080**
1991	8,602	5,814	570	48,250	24,156	63,568	**150,966**
1992	8,102	5,552	533	47,484	24,191	64,597	**150,465**
1993	7,561	5,457	498	48,050	24,560	66,173	**152,305**
1994p	7,850	5,020	520	49,860	24,985	65,270	**153,515**

(1) Includes wood, waste etc from 1990
(2) Includes town gas
(3) Includes solar and geothermal heat from 1990

Table A5
Final consumption of energy by user (heat supplied basis), 1970 to 1994 – United Kingdom

Thousand tonnes of oil equivalent

	Industry	Transport	Domestic	Other[1]	Total
1970	62,192	28,178	36,886	18,729	**146,344**
1975	55,300	30,885	37,062	17,505	**141,417**
1980	48,188	35,541	39,841	18,850	**142,711**
1985	41,611	38,500	42,062	19,694	**142,684**
1990	38,179	48,632	40,725	19,542	**147,080**
1991	37,437	47,970	44,648	20,910	**150,966**
1992	36,242	49,349	44,003	20,872	**150,465**
1993	36,188	50,307	45,405	20,405	**152,305**
1994p	37,055	50,340	45,120	21,000	**153,515**

(1) Mainly agriculture, public administration and commerce

Table A6
Domestic consumption of energy by fuel (heat supplied basis), 1970 to 1994 – United Kingdom

Thousand tonnes of oil equivalent

	Coal	Other solid fuels[1]	Natural gas[2]	Electricity	Petroleum	**Total[3]**
1970	14,242	3,736	8,922	6,622	3,363	**36,886**
1975	8,517	2,423	14,840	7,670	3,612	**37,062**
1980	6,575	1,771	21,258	7,403	2,834	**39,841**
1985	6,290	1,342	24,394	7,582	2,454	**42,062**
1990	3,151	1,197	25,819	8,061	2,491	**40,725**
1991	3,579	1,177	28,704	8,431	2,751	**44,648**
1992	3,106	1,080	28,372	8,549	2,889	**44,003**
1993	3,415	1,078	29,238	8,630	3,038	**45,405**
1994p	3,330	750	29,050	8,770	3,220	**45,120**

(1) Includes wood, waste etc from 1990
(2) Includes town gas
(3) Includes solar and geothermal heat from 1990

Table A7
Industrial consumption of energy by fuel (heat supplied basis), 1970 to 1994 – United Kingdom

Thousand tonnes of oil equivalent

	Coal	Other solid fuels	Coke oven gas[1]	Natural gas[2]	Electricity	Petroleum	**Total[3]**
1970	12,681	9,723	1,164	3,566	6,275	28,782	**62,192**
1975	6,373	6,393	1,038	12,777	6,479	22,241	**55,300**
1980	5,083	3,365	642	15,271	6,854	16,973	**48,188**
1985	4,708	4,715	768	14,868	6,837	9,716	**41,611**
1990	4,170	4,512	602	12,089	8,650	8,157	**38,179**
1991	4,270	4,303	569	10,892	8,557	8,845	**37,437**
1992	4,376	4,220	533	10,581	8,189	8,343	**36,242**
1993	3,558	4,188	498	10,755	8,273	8,915	**36,188**
1994p	3,625	4,170	520	11,840	8,340	8,560	**37,055**

(1) Includes wood, waste etc from 1990
(2) Includes town gas
(3) Includes solar and geothermal heat from 1990

APPENDIX 1

Table A8
Industrial energy consumption by type of use, 1992 – United Kingdom

Percentages

	High temperature processes	Low temperature processes	Compressed air	Motors/ drives	Drying/ separation	Space heating	Refrigeration	Lighting	Other
Iron and steel	87	2	1	3	–	1	–	1	5
Non-ferrous metals[1]	65	–	5	23	–	7	–	–	–
Non-metallic minerals and products	20	–	–	55	20	3	–	2	–
Bricks	75	–	–	5	20	–	–	–	–
Cement, lime and plaster[2]	80	–	–	15	5	–	–	–	–
Glass and glassware	83	–	5	5	–	–	–	–	7
Potteries	55	–	–	15	30	–	–	–	–
Chemicals	–	46	2	10	26	4	3	1	8
Engineering	9	15	10	15	–	48	–	3	–
Vehicles	9	15	10	15	–	50	–	3	–
Food and drink	–	41	2	12	24	13	3	2	3
Textiles, leather and clothing	–	16	1	9	42	29	–	3	–
Paper manufacture and utilisation	–	2	6	40	35	11	3	2	1
Plastics and rubber	–	30	5	35	–	25	2	3	–
Other industries[3]	–	2	7	9	7	65	–	10	–

(1) Lighting is included with space heating
(2) High temperature processes incorporate drying in cement making
(3) Includes timber products and miscellaneous industries
Source: Energy Technology Support Unit

Table A9
Final energy consumption by the service sector by activity and type of fuel, 1991
– United Kingdom

	Electricity	Gas	Petroleum	Solid fuels	Total
		Percentages			Thousand tonnes of oil equivalent
Warehouses	32	35	32	–	1,827
Shops	52	38	7	3	2,893
Hotels and catering	30	59	6	5	3,842
Communication	53	41	6	–	759
Offices	55	40	4	1	1,779
Social	27	52	13	8	1,684
Other commercial	44	33	11	11	190
Government buildings	30	41	25	4	1,731
Education	17	37	39	7	2,063
Health	16	40	32	12	2,774
Total service sector	**33**	**44**	**18**	**5**	**19,542**

Source: Building Research Establishment

APPENDIX 1

Table A10
Coal supply, 1970 to 1994 – United Kingdom

Thousand tonnes

	Production			Imports	Shipments	Stocks[2]
	Total[1]	Deep-mined	Opencast			
1970	**147,195**	136,686	7,885	79	3,191	20,630
1975	**128,683**	117,412	10,414	5,083	2,182	31,358
1980	**130,097**	112,430	15,779	7,334	3,809	37,687
1985	**94,111**	75,289	15,569	12,732	2,432	34,979
1990	**92,762**	72,899	18,134	14,783	2,533	37,760
1991	**94,202**	73,357	18,636	19,611	1,672	43,321
1992	**84,493**	65,800	18,187	20,339	668	47,207
1993	**68,199**	50,457	17,006	18,400	691	45,860
1994p	**48,030**	31,102	16,614	15,850	985	26,572

(1) Includes estimates for slurry etc., recovered and disposed of otherwise than by British Coal.
(2) Total stocks (distributed and undistributed) at end of year.

Table A11
Coal use, 1970 to 1994 – United Kingdom

Thousand tonnes

	Fuel producers		Final consumers			
	Power stations	Other fuel producers	Industry	Domestic	Other final users	Total
1970	77,237	35,686	19,792	20,062	4,108	**43,962**
1975	74,569	24,395	9,678	11,651	1,924	**23,253**
1980	89,569	15,295	7,841	8,946	1,809	**18,596**
1985	73,940	13,630	7,474	8,636	1,707	**17,817**
1990	84,014	12,395	6,283	4,239	1,208	**11,730**
1991	83,542	11,512	6,426	4,778	1,144	**12,348**
1992	78,509	10,350	6,581	4,156	945	**11,682**
1993	66,163	9,807	5,300	4,638	826	**10,765**
1994p	62,572	9,784	5,267	4,052	699	**10,018**

Table A12

Electricity – generation, supply and use, 1970 to 1994[1] – United Kingdom

	Electricity generated	Electricity supplied[2]		Consumption	
		Gross	Net	Fuel producers	Final users
	GWh			TWh	
1970	228,236	212,391	210,904	6.59	192.41
1975	251,263	234,666	233,236	6.29	212.78
1980	266,383	249,120	247,667	6.86	224.25
1985	279,972	261,737	258,242	7.76	234.09
1990	298,495	280,604	277,978	7.27	265.05
1991	301,490	283,066	280,957	7.31	270.44
1992	300,177	281,692	279,435	7.23	271.20
1993	300,514	283,123	281,175	6.64	276.13
1994p	301,130	286,680	284,620	5.95	279.50

(1) Major power producers, transport undertakings and industrial nuclear and hydro stations only, generation and supply are by major power producers only.

(2) The difference between gross and net electricity supplied is the amount of electricity used in pumping at pumped storage stations.

Table A13

Fuel used for electricity generation[1], 1970 to 1994 – United Kingdom

Million tonnes of oil equivalent

	Total all fuels[2]	Coal	Oil	Natural gas	Electricity	
					Nuclear	Natural flow hydro
1970	**68.6**	47.9	13.3	0.1	7.0	0.4
1975	**71.3**	46.9	13.7	2.1	8.1	0.3
1980	**74.9**	56.4	7.7	0.4	9.9	0.3
1985	**75.3**	46.6	11.4	0.4	14.5	0.4
1990	**75.3**	52.0	6.8	–	15.1	0.4
1991	**75.6**	51.7	5.8	0.1	16.3	0.3
1992	**71.9**	46.0	5.4	1.2	17.5	0.4
1993	**70.9**	38.3	4.4	6.3	20.2	0.3
1994p	**70.8**	36.2	3.6	9.1	20.0	0.4

(1) Major power prod}cers, transport undertakings and industrial hydro and nuclear stations only.

(2) Includes imported electricity.

APPENDIX 1

Table A14
Crude oil and petroleum products, 1970 to 1994 – United Kingdom

Thousand tonnes

	Crude oil[1]				Oil products			
	Arrivals	Indigenous production	Shipments	Refinery throughput	Refinery output	Shipments[2]	Arrivals	Inland deliveries
1970	102,155	156	1,182	101,911	94,696	17,424	20,428	91,151
1975	91,366	1,564	1,524	93,597	86,647	13,924	12,786	82,824
1980	46,717	80,467	40,180	86,341	79,227	14,110	9,245	71,177
1985	35,576	127,611	82,980	78,431	72,904	14,828	13,101	69,781
1990	52,710	91,604	56,999	88,692	82,286	16,899	11,005	73,943
1991	57,084	91,260	55,131	92,001	85,476	19,351	10,140	74,506
1992	57,683	94,251	57,627	92,334	85,783	20,250	10,567	75,470
1993	61,701	100,085	64,221	96,274	89,584	23,060	10,064	75,790
1994p	53,096	126,659	82,245	93,162	86,644	22,228	10,190	74,630

(1) Including natural gas liquids and feedstocks.
(2) Shipments of refinery petroleum products.

Table A15
Inland deliveries of petroleum products, 1970 to 1994 – United Kingdom

Thousand tonnes

	Deliveries for energy uses						Deliveries for non-energy uses
	Motor spirit	Derv fuel	Aviation turbine fuel	Gas oil[1]	Fuel oils[2]	**Total[3]**	
1970	14,235	5,035	3,254	11,554	38,585	**81,018**	10,133
1975	16,125	5,414	3,834	12,599	30,470	**73,383**	9,438
1980	19,145	5,854	4,685	11,611	19,157	**64,176**	7,001
1985	20,403	7,106	5,007	9,711	15,969	**61,298**	8,482
1990	24,312	10,652	6,589	8,033	11,997	**64,774**	9,169
1991	24,021	10,694	6,176	8,022	11,948	**64,553**	9,953
1992	24,044	11,132	6,666	7,864	11,481	**64,839**	10,631
1993	23,766	11,806	7,106	7,773	10,770	**65,065**	10,725
1994p	22,834	12,875	7,201	7,488	9,244	**63,592**	11,038

(1) Other than Derv fuel.
(2) Including orimulsion.
(3) Including burning oil, aviation spirit, marine diesel oil, petroleum gases and naphtha (LDF).

Table A16
Gas production and consumption, 1970 to 1994 – Great Britain

GWh

	Production[1]	Arrivals	Shipments	Final consumption[1][2]				
				Total[3]	Domestic	Industry[4]	Electricity generators	Services[5]
1970	171,329	9,759	–	**171,564**	103,806	41,499	1,758	22,332
1975	403,325	9,818	–	**391,250**	172,648	148,646	25,145	39,477
1980	405,346	116,291	–	**508,684**	247,323	177,660	4,103	60,402
1985	462,349	147,122	–	**581,951**	283,810	172,970	5,773	78,162
1990	531,377	79,833	–	**600,085**	300,410	165,002	8,254	87,297
1991	591,794	72,007	–	**643,483**	333,954	158,040	9,429	100,683
1992	602,322	61,255	620	**646,068**	330,100	148,417	22,974	99,253
1993	708,494	48,528	6,824	**721,641**	340,168	148,445	91,744	93,748
1994p	759,876	33,053	9,557	**766,245**	329,541	167,639	119,161	95,802

(1) Includes town gas, colliery methane and substitute natural gas. Includes landfill and sewage gas from 1990. Excludes coke oven gas, blast furnace gas and petroleum gases such as propane.
(2) From 1990 autogenerators are included with electricity generators rather than with industry and services.
(3) Includes oil and gas producers' own use for drilling, production, and pumping operations, and gas industry own use. For 1985 and 1990 to 1994 'other' fuel producers' use (i.e. excluding electricity generators) is included.
(4) Figures for 1970, 1975 and 1980 include 'other' fuel producers.
(5) Includes public administration and agriculture.

Table A17
Energy costs to industry, 1989 – Great Britain

Industry sector	Number of establishments sampled	Total production costs	Total energy costs[1]	Energy costs as a proportion of total production costs
	Number	£ million	£ million	Per cent
Water supply	44	1,450	160	10.8
Paper and board	87	2,540	220	8.8
Bricks, ceramics and refractories	179	1,970	170	8.7
Ferrous foundries	134	1,230	100	8.2
Basic chemicals	68	5,270	420	8.0
Cement, lime and plaster	211	3,910	300	7.7
Extraction of minerals	232	2,420	180	7.6
Glass and glassware	105	1,650	120	7.3
Iron and steel	205	8,260	460	5.6
Non-ferrous foundries	112	570	20	4.3
Non-ferrous metals	135	4,510	170	3.8
Other chemicals	584	17,380	630	3.6
Rubber and plastics	1,073	9,980	310	3.1
Textile manufacture	968	6,890	180	2.6
Food, drink and tobacco	1,416	34,700	870	2.5
Manufacture of metal goods	1,800	9,820	230	2.3
Paper conversion	553	5,830	130	2.2
Other industries	2,185	10,490	180	1.7
Vehicle manufacture	685	28,730	480	1.7
Engineering	4,535	52,680	740	1.4
Printing and publishing	1,717	14,000	140	1.0
Clothing and footwear	1,708	7,090	60	0.9
All industries	**18,733**	**231,380**	**6,280**	**2.7**

(1) Total energy costs are based on fuels purchased and exclude non-traded fuels such as blast furnace gas and coke oven gas.
Source: Energy Paper 64, 'Industrial Energy Markets'

APPENDIX 1

Table A18
Typical retail prices of domestic fuels in certain large cities, 1970 to 1994

		Gas		Electricity	
		Annual level of consumption (kWh)			
		2,500[1]	36,000[2]	2,500[2]	20,000[3]
		Pence per kWh			
Birmingham	December 1970	0.47	0.27	1.00	0.49
	December 1975	0.69	0.45	2.29	1.36
	December 1980	1.16	0.78	4.98	3.30
	December 1985	2.27	1.31	6.49	3.13
	December 1990	2.71	1.62	8.12	3.83
	December 1991	2.79	1.67	9.01	4.22
	December 1992	2.68	1.58	9.10	4.26
	December 1993	2.68	1.58	8.75	4.26
	December 1994[4]	2.90	1.71	9.27	4.59
Edinburgh	December 1970	0.64	0.36	0.96	0.51
	December 1975	0.90	0.55	2.01	1.28
	December 1980	1.37	0.81	5.88	3.11
	December 1985	2.56	1.37	5.81	3.06
	December 1990	2.71	1.62	7.66	3.86
	December 1991	2.79	1.67	8.35	4.22
	December 1992	2.68	1.58	8.51	4.31
	December 1993	2.68	1.58	8.75	4.43
	December 1994[4]	2.90	1.71	9.23	4.81
London	December 1970	0.52	0.31	1.11	0.58
	December 1975	0.88	0.51	2.39	1.50
	December 1980	1.37	0.81	5.54	3.63
	December 1985	2.56	1.37	6.91	3.22
	December 1990	2.71	1.62	8.42	3.87
	December 1991	2.79	1.67	9.42	4.26
	December 1992	2.68	1.58	9.47	4.35
	December 1993	2.68	1.58	9.27	4.29
	December 1994[4]	2.90	1.71	9.76	4.64

(1) Prices are based on the pre-payment tariff and standing charges have been taken into account.
(2) Prices are based on standard domestic tariffs and standing charges have been taken into account.
(3) Up to 1975 the prices are based on a consumption level of 30,000 kWh, including 22,500 kWh at off-peak tariff rates. From 1980 prices are based on 'Economy 7' tariffs ('White Meter' tariffs for Scotland) with night unit provisions of 15,000 kWh.
(4) December 1994 prices include VAT at 8%.

Table A19
Typical retail prices of petroleum products[1], 1970 to 1994 – United Kingdom

Pence per litre

		Motor spirit				Derv fuel	Standard grade burning oil[2]	Gas oil[2]
				Unleaded				
		2 star	4 star	Super	Premium			
1970	January	6.78	7.15	–	–	6.99	2.17	2.02
1975	January	15.62	15.95	–	–	12.21	5.06	5.26
1980	January	25.98	26.39	–	–	27.80	13.07	13.03
1985	January	40.71	41.54	–	–	40.59	21.60	22.62
1990	January	–	40.92	–	38.37	39.21	15.45	15.46
1991	January	–	45.13	44.38	42.14	43.31	17.52	17.13
1992	January	–	46.93	45.57	43.43	43.19	12.47	12.02
1993	January	–	51.27	49.76	47.13	47.05	14.10	13.52
1994	January	–	55.50	54.48	50.83	51.72	12.94	12.72
1995p	January	–	59.48	58.58	53.91	54.25	13.32	13.93

(1) The approximate estimates are generally representative of prices paid (inclusive of taxes) at the pump on or about the 15th of the month. Estimates from 1980 are based on information provided by oil companies.

(2) Typical prices for deliveries of up to 1,000 litres of standard grade burning oil and between 2,000 and 5,000 litres of gas oil. Prior to 1980 prices were for 900 litres of standard grade burning oil and 2,275 litres of has oil, with higher prices for smaller deliveries.

APPENDIX 1

Table A20
Effective rates of duty on principal hydrocarbon oils, 1970 to 1994 – United Kingdom

Pence per litre

Date from which duty effective		Gas for use as road fuel[1][3]	Motor spirit[1][2][4]		Derv fuel[1]	Fuel oil[6]	Gas oil[5][6]	Kerosene[6]
			Unleaded	Leaded				
15 April	1969	–	–	4.95	4.95	0.22	0.22	0.22
3 July	1972	2.47	–					
10 April	1976	3.30	–	6.60	6.60			
30 March	1977	3.85	–	7.70	7.70	0.55	0.55	
8 August	1977	3.30	–	6.60				
13 June	1979	4.05	–	8.10	9.20	0.66	0.66	
26 March	1980	5.00	–	10.00	10.00	0.77	0.77	
10 March	1981	6.91	–	13.82	13.82			
2 July	1981			–	11.91			
9 March	1982	7.77	–	15.54	13.25			
15 March	1983	8.15	–	16.30	13.82			
13 March	1984	8.58	–	17.16	14.48			zero
19 March	1985	8.97	–	17.94	15.15			
19 March	1986	9.69	–	19.38	16.39		1.10	
17 March	1987		18.42					
15 March	1988	10.22		20.44	17.29			
14 March	1989		17.72					
20 March	1990	11.24	19.49	22.48	19.02	0.83	1.18	
19 March	1991	12.93	22.41	25.85	21.87	0.91	1.29	
10 March	1992	13.90	23.42	27.79	22.85	0.95	1.35	
16 March	1993	15.29	25.76	30.58	25.14	1.05	1.49	
30 November	1993	16.57	28.32	33.14	27.70	1.16	1.64	
29 November	1994	33.14p/kg	30.44	35.26	30.44	1.66	2.14	
1 January	1995	33.14p/kg	31.32	36.14	31.32	1.66	2.14	

(1) These fuels became liable to Value Added Tax as follows:–
   (i) 10% with effect from 1 April 1974
   (ii) 8% with effect from 29 July 1974
   (iii) For motor spirit 25% with effect from 18 November 1974
   (iv) For motor spirit 12.5% with effect from 12 April 1976
   (v) 15% with effect from 18 June 1979
   (vi) 17.5% with effect from 1 April 1991
(2) With effect from 14 March 1989 until 20 March 1990, the rate of duty from 2-star and 3-star leaded motor spirit was 21.220 pence per litre.
(3) Including aviation spirit from 9 March 1982.
(4) Including aviation spirit up to 8 March 1982.
(5) AVTUR (aviation turbine fuel) attracted the gas oil rate until 18 March 1986, after which it was zero rated.
(6) For industrial and commercial consumers these fuels became liable to the standard rate of Value Added Tax on 1 July 1990 (at 15% to 31 March 1991 and at 17.5% from 1 April 1991), recoverable by the majority of such consumers. For domestic consumers from 1 April 1994 these fuels became liable to Value Added Tax at the rate of 8%.

Table A21
Industrial fuel prices[1], 1970 to 1994 – Great Britain

In original units

	Size of consumer	1970[2]	1975	1980	1985	1988 Old basis[3]
Coal[4]	Small	..	..	..	..	..
(£ per tonne)	Large	..	..	..	..	..
All consumers –	average	6.6	14.6	35.0	51.00	43.10
Heavy fuel oil[5]	Small	..	..	..	..	..
(£ per tonne)	Large	..	..	..	..	..
All consumers –	average	9.2	37.7	90.3	151.75	55.04
Gas oil[5]	Small	..	..	..	..	..
(£ per tonne)	Large	..	..	..	..	..
All consumers –	average	..	52.7	150.6	219.78	107.78
Electricity	Small	..	..	..	..	..
(Pence per kWh)	Large	..	..	..	..	..
All consumers –	average	0.65	1.24	2.37	3.00	3.00
Gas	Small	..	..	..	..	..
(Pence per kWh)	Large	..	..	..	..	..
All consumers –	average	0.15	0.15	0.60	0.97	0.74

	Size of consumer	1988 New basis[3]	1991	1992	1993	1994p
Coal[4]	Small	68.17	67.69	69.67	69.72	63.58
(£ per tonne)	Large	42.63	41.53	40.99	37.60	37.30
All consumers –	average	44.76	43.87	43.48	40.61	40.18
Heavy fuel oil[5]	Small	75.65	74.07	68.68	77.59	79.12
(£ per tonne)	Large	57.20	63.47	62.03	66.21	71.44
All consumers –	average	62.40	66.90	64.34	68.58	74.07
Gas oil[5]	Small	126.90	169.64	149.27	159.00	154.83
(£ per tonne)	Large	108.42	148.03	130.93	139.25	128.13
All consumers –	average	109.87	150.81	132.79	141.64	131.04
Electricity	Small	5.15	6.74	7.06	6.88	6.62
(Pence per kWh)	Large	3.05	3.33	3.57	3.81	3.74
All consumers –	average	3.47	3.83	4.06	4.26	4.15
Gas	Small	1.19	1.36	1.38	1.28	1.72
(Pence per kWh)	Large	0.72	0.69	0.70	0.71	0.73
All consumers –	average	0.78	0.75	0.76	0.77	0.77

(1) Prices of fuels purchased by manufacturing industry.
(2) The figures for 1970 came from a variety of sources, and were merely representative of prices paid at the time.
(3) Between 1975 and 1988 (the 'old basis') the prices were based on a survey of between 800 and 900 large industrial consumers.
From 1988 ('new basis') the survey has been of about 1,200 establishments, grouped according to the approximate level of purchases.
(4) Excludes blast furnace supplies.
(5) Oil product prices include hydrocarbon oil duty.

APPENDIX 1

Table A22
Energy production and consumption in the European Union, 1992

	Production	Imports	Exports	Gross inland consumption	Final consumption	Arrival dependency[1]	Energy ratio[2]
	Million tonnes of oil equivalent					Per cent	Toe per $1,000 GDP
Belgium	11.2	65.1	21.6	50.7	33.0	86.0	0.53
Denmark	11.5	17.9	10.1	18.1	13.7	43.3	0.28
France	101.4	144.3	20.3	221.8	132.2	55.9	0.36
Germany	155.1	206.1	19.8	332.7	181.6	56.0	0.40
Greece	8.0	23.1	5.4	22.1	13.9	80.4	0.59
Ireland	3.2	7.5	0.8	10.0	7.0	67.2	0.39
Italy	25.5	155.8	21.7	155.3	107.8	86.4	0.31
Luxembourg	–	3.8	–	3.8	3.5	99.9	0.83
Netherlands	66.5	104.2	90.3	68.8	44.5	20.3	0.45
Portugal	0.5	19.9	3.4	16.5	10.4	100.2	0.62
Spain	28.8	79.5	12.5	91.7	57.3	73.1	0.43
United Kingdom	209.8	88.1	79.7	214.2	141.0	3.9	0.41

(1) (Imports-exports)/gross inland consumption.
(2) Gross inland consumption/GDP where GDP is expressed in billion US dollars at 1985 prices.
Source: Eurostat

Table A23
Energy production and consumption in some OECD countries, 1992

	Production	Imports	Exports	Total primary energy supply	Final consumption	Arrival dependency[1]	Energy ratio[2]
	Million tonnes of oil equivalent					Per cent	Toe per $1,000 GDP
Australia	171.4	15.6	97.3	88.8	59.6	–92.1	0.46
Austria	8.7	19.1	1.2	25.9	22.4	69.1	0.33
Canada	295.1	44.0	127.7	216.3	163.4	–38.7	0.55
Finland	11.7	20.0	4.0	28.0	23.1	57.1	0.50
Japan	74.0	389.8	7.6	451.1	319.4	84.7	0.26
New Zealand	12.9	3.8	1.7	14.7	10.4	14.1	0.63
Norway	145.4	4.6	127.7	21.3	17.8	–578.7	0.32
Sweden	29.1	28.6	10.8	46.7	32.8	38.1	0.43
Switzerland	9.8	17.7	2.5	25.3	20.7	60.0	0.24
Turkey	27.0	29.8	1.9	55.5	43.7	50.3	0.73
United Kingdom[3]	213.4	87.7	80.8	216.2	152.1	3.2	0.41
United States	1,658.8	466.8	116.1	1,984.1	1,399.4	17.7	0.43

(1) (Imports-exports)/total primary energy supply.
(2) Total primary energy supply/GDP where GDP is expressed in billion US dollars at 1985 prices.
(3) The figures for the UK differ slightly from those published by Eurostat (Table A22) as a result of methodological differences.
Source: International Energy Agency

Table A24

Electricity and gas prices in European Union and some OECD countries in 1993

US cents per kWh

	Electricity		Natural Gas[1]	
	Industry	Households	Industry	Households
EU				
Belgium	5.96	16.75	1.11	3.42
Denmark	7.04	17.99	..	5.52
France	5.46	14.62	1.23	3.92
Germany	8.90	16.83	1.60	3.64
Greece	5.93	10.22	..	..
Ireland	5.93	11.96	2.31	3.68
Italy	9.25	14.60	1.29	5.47
Luxembourg	6.01[2]	11.46	..	2.14
Netherlands	5.29[3]	11.91[4]	1.08[3]	2.70
Portugal	12.14	16.36	..	..
Spain	8.47	17.68	1.31[4]	4.17[2]
United Kingdom	6.73	11.25	1.10	2.41
OECD				
Australia	4.19	7.07	1.04	2.37
Austria	7.14	16.34	1.46	3.41
Canada	3.92	6.26	0.64	1.52
Finland	4.84	8.05	0.91	1.11
Japan	16.19	23.11	4.00	10.35
New Zealand	3.31	5.84	1.58	2.19
Norway	3.70[3]	6.77	..	..
Sweden	3.53	8.19	..	..
Switzerland	9.53	11.86	2.28	3.68
Turkey	9.48	9.86	1.34	2.18
United States	4.87	8.33	1.02	2.04

Note: Prices include taxes except where refundable.
(1) The prices for natural gas refer to kWh using the gross calorific value of gas. Prices for kWh using the net calorific value would be approximately 11% higher.
(2) Latest available data are for 1989.
(3) Latest available data are for 1991.
(4) Latest available data are for 1992.
Source: IEA Energy Prices and Taxes.

APPENDIX 2 ENERGY RELATED ATMOSPHERIC EMISSIONS

A2.1 The operations of the energy sector in the UK, as elsewhere, can affect the environment in many different ways. Detrimental effects can result from exploration, production, transportation, storage, conversion and distribution. The final use of the energy and the disposal of waste products can also damage the environment.

A2.2 The particular areas of potential environmental concern related to the energy sector are:-

ambient air quality; acid deposition; coal mining subsidence; major environmental accidents; water pollution; maritime pollution; land use and siting impact; radiation and radioactivity; solid waste disposal; hazardous air pollutants; stratospheric ozone depletion; and global climate change.

A2.3 There has been an increasing awareness of the environmental damage resulting from energy activities in recent years. In a

Table A2.1

Importance of fossil fuel use in the generation of air pollutants in 1993

Pollutant	Fossil fuel emissions as a percentage of total man-made emissions	Contribution of different fossil fuels as a percentage of total man-made emissions		Contribution of different sources in total of man-made emissions	
Carbon dioxide	97%	solid fuels:	36%	power stations:	30%
		oil products:	37%	other industry[1]:	22%
		gas:	24%	transport;	22%
Methane	24%	coal:	12%	landfill:	46%
		oil and gas:	12%	livestock:	27%
Sulphur dioxide	99%	solid fuels:	74%	power stations	65%
		oil products:	25%	other industry[1]:	19%
Black smoke	87%	solid fuels:	33%	transport[2]:	49%
		oil products:	54%	domestic:	27%
Oxides of nitrogen	95%	solid fuels:	25%	transport[2]:	56%
		oil products:	62%	power stations:	24%
		gas:	7%		
Volatile organic compounds	97%	solid fuels:	2%	industrial processes and solvents:	50%
		oil products and gas:	95%	transport[2]:	38%
Carbon monoxide	95%	solid fuels:	6%	transport:	89%
		oil products:	89%		

[1] Excludes power stations, refineries and agriculture
[2] Includes only a restricted amount of emissions from civil aircraft and shipping.

survey in 1993, commissioned by the Department of the Environment, of more than three thousand adults in England and Wales, 60 per cent of respondents were very worried about radioactive waste, 52 per cent about oil spills at sea, 40 per cent about traffic exhausts and smog, 35 per cent about global warming, and 31 per cent about acid rain.

A2.4 This appendix brings together the latest data on air pollution arising, at least in part, from energy related activities, updating the data presented in Annex D of the 1994 *Digest of UK Energy Statistics*. Figures are based on data from the National Environmental Technology Centre, which are presented in further detail in the Department of the Environment's *Digest of Environmental Protection and Water Statistics 1995*.

A2.5 Table A2.1 shows the importance of energy activities in the generation of seven major pollutants. The remainder of this appendix looks in more detail at the latest information available on each of these pollutants.

GREENHOUSE GASES

A2.6 The greenhouse gases maintain the earth's surface at 33°C higher than it would be in their absence. Water vapour is by far the most important natural greenhouse gas but its amount is largely determined by atmospheric temperature and the amount of heating of the earth's surface. Carbon dioxide is also an important natural greenhouse gas, and some methane is natural. Human action is increasing the concentrations in the atmosphere of carbon dioxide, methane and nitrous oxide, and is wholly responsible for halocarbons, of which the chlorofluorocarbons are the most significant. The latter may, however, have indirect effects which significantly offset their warming potential.

Carbon dioxide

A2.7 The most important man-made greenhouse gas is carbon dioxide. Currently world-wide emissions of carbon dioxide are responsible for 62 per cent of global warming. Although this gas is naturally emitted by living organisms, these emissions are balanced by the uptake of carbon dioxide by plants during photosynthesis; they therefore tend to have no net effect on atmospheric concentrations and can be ignored. The burning of fossil fuels, however, releases carbon dioxide fixed by plants many millions of years ago, increasing its concentration in the atmosphere and reducing the heat lost from the atmosphere into space.

A2.8 The UK contributes about 2 per cent of global emissions of carbon dioxide; the UK's emissions are broken down by fuel and by source in Table A2.2. In 1993 over 97 per cent of carbon dioxide emissions came directly from fossil fuel combustion: coal (34 per cent), natural gas (24 per cent) and motor spirit (13 per cent). Three per cent of emissions came from non-energy products. The main sources of emissions were power stations (30 per cent), other industries (22 per cent), road transport (20 per cent) and the domestic sector (16 per cent). Emissions from power stations fell in 1993 for the third year in succession, as the switch from coal to gas continued and as the contribution of nuclear power grew. Trends in carbon dioxide emissions since 1970 are shown in Chart A2.1.

Table A2.2

Carbon dioxide emissions in 1993

	Million tonnes of carbon	
By fuel		
Coal	52	(34%)
Solid smokeless fuel	2	(2%)
Petroleum:		
Motor spirit	20	(13%)
DERV	10	(7%)
Gas oil	7	(5%)
Fuel oil	12	(8%)
Burning oil	2	(1%)
Other petroleum	5	(3%)
Natural gas	36	(24%)
Other emissions	5	(3%)
Total	**152**	**(100%)**
By source		
Domestic sector	24	(16%)
Commercial sector	8	(5%)
Power stations	46	(30%)
Refineries	6	(4%)
Other industry	34	(22%)
Road transport	30	(20%)
Other transport	3	(2%)
Agriculture	1	(-)
Total	**152**	**(100%)**

APPENDIX 2

Chart A2.1

Carbon dioxide emissions[1] by source, 1970 to 1993.

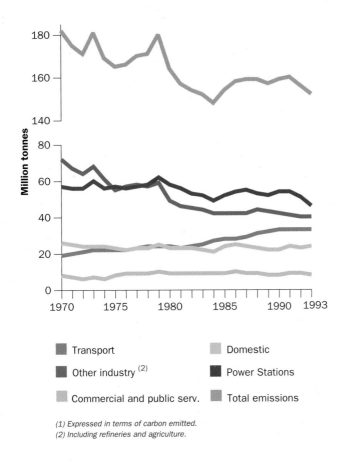

- Transport
- Other industry [2]
- Commercial and public serv.
- Domestic
- Power Stations
- Total emissions

(1) Expressed in terms of carbon emitted.
(2) Including refineries and agriculture.

Source: Department of Trade and Industry

A2.9 In 1993 emissions of carbon dioxide were virtually the same as in 1983, and 17 per cent lower than in 1970, when more was emitted by power stations and industry. Levels typically fluctuated between 150 and 160 million tonnes throughout the 1980s with the exception of the relatively low figure of 148 million tonnes in 1984 which was a result of the miners' strike.

Methane

A2.10 Methane is the second most important man-made greenhouse gas. Like carbon dioxide it also occurs naturally as a component of biological cycles. Currently, emissions world-wide are responsible for 20 per cent of global warming. Concentrations in the atmosphere have been rising at a rate of 0.8 per cent per annum over the last few decades, although there has been a slow down in the rate of increase in recent years. Table A2.3 shows UK methane emissions in 1993 by source. The main sources were livestock (27 per cent), landfill waste disposal sites (46 per cent) and coal mining (12 per cent). Unlike carbon dioxide only about 1 per cent came from fuel combustion.

A2.11 In 1993 methane emissions were 15 per cent lower than in 1983, and also 15 per cent lower than in 1970. Since 1986 emissions have generally fallen steadily, with reductions from coal mining, domestic and commercial combustion and cattle.

A2.12 It is, however, difficult to obtain reliable estimates for methane. Figures for emissions from landfill sites and livestock are very uncertain. The number of significant figures quoted in this table should not be taken as an indication of the accuracy of the estimate.

Table A2.3
Methane emissions in 1993

	Thousand tonnes	
By source		
Domestic and commercial combustion	49	(1%)
Power stations	3	(-)
Deep mined coal	516	(12%)
Open cast coal	5	(-)
Offshore oil and gas	111	(3%)
Gas leakage	370	(9%)
Industrial combustion	8	(-)
Landfill	1,918	(46%)
Sewage disposal	72	(2%)
Road transport	11	(-)
Cattle	794	(19%)
Sheep	264	(7%)
Other animals	51	(1%)
Total	**4,173**	**(100%)**

AIR POLLUTION

A2.13 Air pollution not only damages human health but also has a wide range of environmental impacts, affecting soil, water, wildlife, crops, forests and buildings. This section gives information on the main air pollutants associated with fossil fuel combustion - sulphur dioxide, black smoke, nitrogen oxides, volatile organic compounds and carbon monoxide.

Sulphur dioxide

A2.14 Sulphur dioxide is a gas produced by the combustion of sulphur-containing fuels such as coal and oil. It is one of the main gases responsible for acid deposition. Table A2.4 shows figures for sulphur dioxide emissions in 1993 in the UK by fuel and by source. Most sulphur dioxide is generated by the combustion of coal. In 1993 65 per cent of sulphur dioxide was emitted by power stations, with a further 24 per cent coming from refineries and other industrial sources.

Table A2.4
Sulphur dioxide emissions in 1993

	Thousand tonnes	
By fuel		
Coal	2,303	(72%)
Solid smokeless fuel	50	(2%)
Petroleum:		
Motor spirit	14	(-)
DERV	45	(1%)
Gas oil	53	(2%)
Fuel oil	677	(21%)
Burning oil	2	(-)
Other petroleum	27	(1%)
Other emissions	24	(1%)
Total	**3,194**	**(100%)**
By source		
Domestic sector	113	(4%)
Commercial sector	88	(3%)
Power stations	2,089	(65%)
Refineries	156	(5%)
Other industry	625	(19%)
Road transport	59	(2%)
Other transport	55	(2%)
Agriculture	10	(-)
Total	**3,194**	**(100%)**

Chart A2.2
Sulphur dioxide emissions by sector, 1970 to 1993.

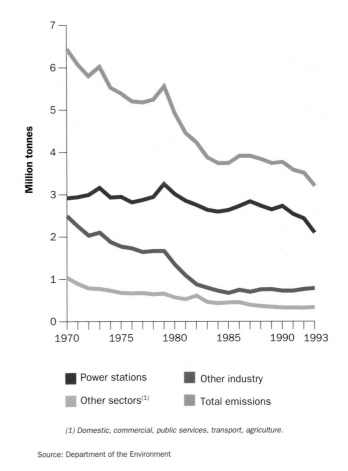

(1) Domestic, commercial, public services, transport, agriculture.

Source: Department of the Environment

A2.15 Trends in sulphur dioxide emissions since 1970 are shown in Chart A2.2. In 1993 sulphur dioxide emissions were 18 per cent lower than in 1983 and 50 per cent lower than in 1970, as a result of lower coal and fuel oil consumption. The reduction in the quantity of coal used by power stations has had a marked effect in recent years.

Black smoke

A2.16 Black smoke consists of fine suspended particles from incomplete fuel combustion and volatilisation of trace elements present in fuel. Table A2.5 shows black smoke emissions by fuel and by source in the UK. In 1993 road transport generated almost a half of all emissions, mostly from the combustion of DERV, whilst the domestic sector was responsible for 27 per cent.

A2.17 In 1993 black smoke emissions were around 9 per cent lower than in 1983 and 55 per cent lower than in 1970. Emissions by the domestic sector have fallen by 84 per cent since 1970, while emissions from road transport have risen by 128 per cent over the same period.

Nitrogen oxides

A2.18 A number of nitrogen compounds including nitrogen dioxide, nitric oxide and nitrous oxide are formed in combustion processes when nitrogen in the air or the fuel combines with oxygen. Along with sulphur dioxide they are responsible for acid deposition, while nitrous oxide is a greenhouse gas. Table A2.6 shows figures for emissions of nitrogen oxides in 1993 by fuel and by source. Almost a half of all emissions of nitrogen oxides came from road transport, with a further 24 per cent coming from power stations.

A2.19 Trends in nitrogen oxides emissions since 1970 are shown in Chart A2.3. Emissions rose steadily between 1985 and 1990, as increasing amounts were generated by road transport. Since 1990 emissions from road transport have declined with the introduction of catalytic converters, whilst emissions

Table A2.5

Black smoke emissions in 1993

	Thousand tonnes	
By fuel		
Coal	136	(29%)
Solid smokeless fuel	17	(4%)
Petroleum:		
Motor spirit	15	(3%)
DERV	213	(46%)
Gas oil	7	(1%)
Fuel oil	14	(3%)
Other petroleum	1	(-)
Other emissions	60	(13%)
Total	**463**	**(100%)**
By source		
Domestic sector	127	(27%)
Commercial sector	4	(1%)
Power stations	21	(4%)
Refineries	3	(1%)
Other industry	77	(17%)
Road transport	228	(49%)
Other transport	4	(1%)
Agriculture	1	(-)
Total	**463**	**(100%)**

APPENDIX 2

Table A2.6
Emissions of nitrogen oxides in 1993

	Thousand tonnes	
By fuel		
Coal	592	(25%)
Solid smokeless fuel	5	(-)
Petroleum:		
Motor spirit	714	(30%)
DERV	431	(18%)
Gas oil	152	(6%)
Fuel oil	139	(6%)
Burning oil	6	(-)
Other petroleum	41	(2%)
Natural gas	156	(7%)
Other emissions	119	(5%)
Total	**2,355**	**(100%)**
By source		
Domestic sector	76	(3%)
Commercial sector	55	(2%)
Power stations	570	(24%)
Refineries	40	(2%)
Extraction anddistribution		
of fossil fuels	98	(4%)
Other industry	202	(9%)
Road transport	1,144	(49%)
Other transport	164	(7%)
Agriculture	4	(-)
Total	**2,355**	**(100%)**

Chart A2.3
Nitrogen oxides emissions[1] by sector, 1970 to 1993.

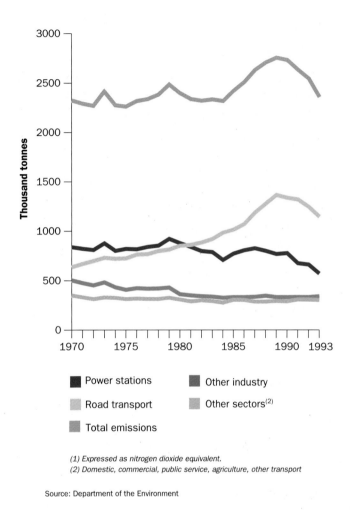

(1) Expressed as nitrogen dioxide equivalent.
(2) Domestic, commercial, public service, agriculture, other transport

Source: Department of the Environment

from power stations have declined with the reduction in the use of coal. The level of emissions in 1993 was 10 per cent higher than in 1983.

Volatile organic compounds

A2.20 Volatile organic compounds consist of a wide range of chemicals, including hydrocarbons and oxygenated and halogenated organics, which are released from oil refining, petrol distribution, motor vehicles, industrial processes and solvent use. Apart from methane (which is dealt with separately above) the main environmental effect of these compounds is through their involvement in the creation of ground level ozone. In addition benzene is a known carcinogen and 1,3 butadiene is a suspected carcinogen.

A2.21 Table A2.7 shows emissions of volatile organic compounds in 1993 by fuel and by source. More than half these emissions were produced during the extraction, distribution or combustion of fossil fuels, with further emissions resulting from the use of solvents of fossil fuel origin.

A2.22 Figures for 1993 show that emissions of volatile organic compounds were 3 per cent higher than in 1983 and 12 per cent higher than in 1970. Over the last two decades declining emissions from the domestic sector have been more than offset by increases from road transport and offshore oil and gas activity.

Carbon monoxide

A2.23 Carbon monoxide is derived from the incomplete combustion of fuel, mainly from road transport. Table A2.8 gives data for carbon monoxide emissions by source. In 1993 road transport accounted for 88 per cent of these emissions. Trends in carbon monoxide emissions are shown in Chart A2.4. Emissions from road transport have declined since 1991 with the introduction of catalytic converters. In 1993 carbon monoxide emissions were 15 per cent higher than in 1983 and 28 per cent higher than in 1970, reflecting increases in road transport.

Table A2.7
Emissions of volatile organic compounds in 1993

	Thousand tonnes	
By fuel		
Coal	38	(2%)
Solid smokeless fuel	4	(-)
Petroleum:		
Motor spirit	696	(28%)
DERV	77	(3%)
Gas oil	19	(1%)
Fuel oil	3	(-)
Other petroleum	5	(-)
Natural gas	6	(-)
Other emissions[1]	1,606	(65%)
Total	**2,453**	**(100%)**
By source		
Domestic sector	35	(1%)
Commercial sector	1	(-)
Power stations	10	(-)
Processes and solvents	1,222	(50%)
Other industry	166	(7%)
Road transport	913	(37%)
Other transport	25	(1%)
Forestry	80	(3%)
Total	**2,453**	**(100%)**

(1) Includes industrial processes and solvents, petrol evaporation, offshore oil and gas activities, gas leakage and forestry.

Table A2.8
Carbon monoxide emissions in 1993

	Thousand tonnes	
By fuel		
Coal	226	(4%)
Solid smokeless fuel	138	(2%)
Petroleum:		
Motor spirit	4,942	(85%)
ERV	178	(3%)
Gas oil	27	(-)
Fuel oil	14	(-)
Other petroleum	12	(-)
Natural gas	12	(-)
Other emissions	262	(5%)
Total	**5,813**	**(100%)**
By source		
Domestic sector	279	(5%)
Commercial sector	6	(-)
Power stations	38	(1%)
Refineries	2	(-)
Other industry	326	(6%)
Road transport	5,121	(88%)
Other transport	42	(1%)
Agriculture	1	(-)
Total	**5,813**	**(100%)**

Chart A2.4
Carbon monoxide emissions by sector, 1970 to 1993.

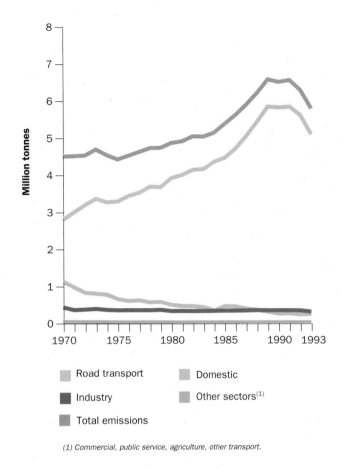

Road transport — Domestic
Industry — Other sectors[1]
Total emissions

(1) Commercial, public service, agriculture, other transport.

Source: Department of the Environment

APPENDIX 3 POWER STATIONS, OIL REFINERIES, AND COAL MINES

Power Stations in the United Kingdom

Company Name	Station Name	Fuel	Capacity (MW)	Year of Commission
National Grid Company	Dinorwig	pumped storage	1,740	1983
	Ffestiniog	pumped storage	360	1961
National Power	Aberthaw A (mothballed)	coal	0	1960
	Aberthaw B	coal	1,455	1971
	Blyth A	coal	456	1958
	Blyth B	coal	626	1965
	Didcot	coal	1,960	1972
	Drax	coal	3,870	1974
	Eggborough	coal	1,980	1968
	Ironbridge	coal	970	1970
	Rugeley A (mothballed)	coal	0	1961
	Rugeley B	coal	976	1972
	Skelton Grange (mothballed)	coal	0	1961
	Uskmouth (mothballed)	coal	0	1962
	West Burton	coal	1,957	1967
	Willington A (mothballed)	coal	0	1957
	Willington B	coal	376	1962
	Fawley	oil	484	1969
	Littlebrook D	oil	1,370	1982
	Pembroke	oil	974	1970
	Tilbury B	coal/oil	1,360	1965
	Aberthaw GT	gas oil	51	1967
	Cowes GT	gas oil	140	1982
	Didcot GT	gas oil	100	1968
	Drax GT	gas oil	105	1971
	Eggborough GT (mothballed)	gas oil	0	1967
	Fawley GT	gas oil	68	1969
	Ironbridge GT (mothballed)	gas oil	0	1967
	Letchworth GT	gas oil	140	1979
	Littlebrook GT	gas oil	105	1980
	Norwich GT	gas oil	110	1966
	Ocker Hill GT	gas oil	280	1979
	Pembroke GT (mothballed)	gas oil	0	1969
	Rugeley GT	gas oil	50	1969
	Tilbury GT (mothballed)	gas oil	0	1965
	West Burton GT	gas oil	40	1966
	Deeside	CCGT	528	1994
	Killingholme	CCGT	620	1993
	Little Barford	CCGT	750	1994
Nuclear Electric	Maentwrog	hydro	30	1928
	Bradwell	Magnox	245	1962
	Dungeness A	Magnox	440	1965
	Hinkley Point A	Magnox	470	1965
	Oldbury	Magnox	434	1967
	Sizewell A	Magnox	420	1966
	Wylfa	Magnox	950	1971
	Dungeness B	AGR	1,110	1985
	Hartlepool	AGR	1,210	1984
	Heysham 1	AGR	1,150	1984
	Heysham 2	AGR	1,250	1988
	Hinkley Point B	AGR	1,195	1978
	Sizewell B	PWR	1,188	1995
PowerGen	Cottam	coal	1,988	1969
	Drakelow C	coal	960	1965
	Ferrybridge C	coal	1,960	1966
	Fiddler's Ferry	coal	1,455	1971
	High Marnham	coal	945	1959
	Kingsnorth	coal	1,940	1970
	Ratcliffe	coal	1,995	1968
	Grain	oil	2,025	1979
	Richborough	orimulsion	230	1962
	Ince B	oil/orimulsion	480	1982
	Ferrybridge GT	gas oil	34	1966
	Fiddler's Ferry GT	gas oil	34	1969
	Grain GT	gas oil	58	1978
	Ince GT	gas oil	50	1979
	Kingsnorth GT	gas oil	34	1967
	Ratcliffe GT	gas oil	34	1966
	Taylor's Lane GT	gas oil	132	1979
	Richborough, Kent	wind	1	1990
	Killingholme PG1	CCGT	900	1992
	Rye House	CCGT	740	1993

APPENDIX 3

Company Name	Station Name	Fuel	Capacity (MW)	Year of Commission
Scottish Hydro-Electric	Peterhead	oil/gas	1,550	1980
Schemes:				
Affric/Beauly	Mullardoch	hydro	2.4	
	Fasnakyle	hydro	66	
	Deanie	hydro	38	completed
	Culligran	hydro	24	1963
	Aigas	hydro	20	
	Kilmorack	hydro	20	
Breadalbane	Lubreoch	hydro	4	
	Cashlie	hydro	11	
	Lochay	hydro	47	completed
	Finlarig	hydro	30	1961
	Lednock	hydro	3	
	St. Fillans	hydro	21	
	Dalchonzie	hydro	4	
Conon	Achanalt	hydro	3	
	Grudie Bridge	hydro	24	
	Mossford	hydro	24	completed
	Luichart	hydro	34	1961
	Orrin	hydro	18	
	Torr Achilty	hydro	15	
Foyers	Foyers	hydro	300	completed
	Foyers Falls	hydro	2	1975
	Mucomir	hydro	2	
Garry/Morrison	Ceannacroc	hydro	20	
	Livishie	hydro	15	
	Glenmoriston	hydro	36	completed
	Quoich	hydro	22	1962
	Ivergarry	hydro	20	
Shin	Cassley	hydro	10	completed
	Lairg	hydro	3.5	1960
	Shin	hydro	24	
Sloy/Awe	Sloy	hydro	130	
	Sron Mor	hydro	5	
	Clachan	hydro	40	
	Allt-na-lairige	hydro	6	
	Nant	hydro	15	completed
	Inverawe	hydro	25	1959
	Kilmelfort	hydro	2	
	Loch Gair	hydro	6	
	Lussa	hydro	2.4	
	Striven	hydro	8	
Tummel	Gaur	hydro	6.4	
	Cuaich	hydro	2.5	
	Loch Ericht	hydro	2.2	
	Rannoch	hydro	48	completed
	Tummel	hydro	34	1951
	Errochty	hydro	75	
	Trinafour	hydro	0.5	
	Clunie	hydro	61.2	
	Pitlochry	hydro	15	
Scottish Nuclear	Hunterston B	AGR	1,150	1976
	Torness	AGR	1,250	1988
Scottish Power	Cockenzie	coal	1,152	1967
	Kincardine	coal	375	1962
	Longannet	coal	2,304	1970
	Methil	coal	57	1965
	Clyde's Mill	gas oil	55	1965
	Galloway (6 stations)	hydro	109	1935
	Lanark (2 stations)	hydro	16	1927
	Cruachan	pumped storage	400	1966

Independent Power Stations

Company Name	Station Name	Fuel	Capacity (MW)	Year of Commission
Ballylumford Power Ltd	Ballylumford	oil	1,080	1968
Barking Power	Barking	CCGT	1,000	1995
BFI Packington	Old Railway Cutting	landfill gas	5	1993
Blue Circle Industries	Ockendon Landfill	landfill gas	4	1993
	Beddingham Landfill	landfill gas	2	1993
	Stone Pit 1 & 2 Landfill	landfill gas	3	1992
British Coal Corporation	Nottingham Refuse	waste	7	1992
British Nuclear Fuels	Calder Hall	Magnox	198	1956
	Chapelcross	Magnox	196	1959
CeltPower Ltd	Penrhyddlan wind farm	wind	5	1993
	Llidiartywaun wind farm	wind	7	1993
Citigen Ltd	Charterhouse St., London	gas oil	60	1994
Coolkeeragh Power	Coolkeeragh	oil	360	1959
Corby Power	Corby	CCGT	406	1993
Coventry City Council	Coventry	waste	8	1992
CRE Energy Ltd	Corkey	wind	2	1994
	Rigged Hill	wind	2	1994
Derwent Cogeneration	Spondon	CCGT	236	1995
Ecogen Ltd	St Breock	wind	2	1994
Elm Energy UK	Wolverhampton	waste	20	1994
Fellside Power	Sellafield	CCGT	168	1993
Fibrogen	Glanford	chicken litter	13	1993
Fibropower Ltd	Eye, Suffolk	chicken litter	14	1992
Keadby Developments	Keadby 1	CCGT	750	1995
Lakeland Power Ltd	Roosecote	CCGT	229	1991
National Wind Power	Llangwyryfon wind farm	wind	6	1993
	Camarthen Bay	wind	1	1986
	Kirby Moor, Ulverston	wind	5	1993
	Cemmaes Valley, Powys	wind	7	1992
	Cold Northcott wind farm	wind	7	1993
Nigen Ltd	Belfast West	coal	240	1954
	Kilroot	coal/oil	580	1981
North London Waste Auth.	Edmonton	waste	27	1990
North West Water	Davyhulme, Urmaston	sewage gas	5	1992
	5 other small schemes	sewage gas	4	1992
Peterborough Power	Peterborough	CCGT	405	1993
Regional Power Gen.	Brigg	CCGT	272	1993

APPENDIX 3

Company Name	Station Name	Fuel	Capacity (MW)	Year of Commission
SELCHEP	Kennels Site	waste	25	1994
Shanks & McEwan (Southern)	Brogborough I & II	landfill gas	11	1992
	Calvert	landfill gas	3	1992
Slough Estates	Edinburgh Avenue	waste	12	1994
Southampton Geothermal	Southampton	waste	5	1993
Teesside Power Ltd	Teesside	CCGT	1,845	1992
Thames Water	Beckton, East London	sewage gas	8	1993
	Mogden, Isleworth	sewage gas	4	1993
	12 other small schemes	sewage gas	7	1992
UK Windfarms Ltd	Marcheini	wind	4	1994
Vestas DWT	Caton Moor	wind	2	1994
West Coast Windfarms	Fullabrook/Crackway	wind	5	1993
Wind Power Systems Ltd	Dyffryn-Brodyn	wind	2	1994
Wind Resources Ltd	Carland Cross, Mitchell	wind	6	1992
	Coal Clough	wind	10	1993

Sources include: Electricity Association, National Grid Company, the DTI and other sources.

This list covers stations of more than 1 MW capacity. A number of smaller renewables schemes have been commissioned under the Non-Fossil Fuel Obligation (NFFO). The table below shows the overall capacity commissioned under the first two NFFO Orders for England and Wales; the figures include capacity at the larger stations mentioned in the list above.
The Third NFFO Order for England and Wales is not included because the public electricity suppliers were not required to comply with it until 20 January 1995. It is also too soon to include the First Scottish Renewables Order. The First NFFO Order for Northern Ireland was announced in April 1994 and the two largest stations to have been commissioned at the time of printing are Corkey and Rigged Hill wind stations, which are both mentioned in the above list.

NFFO Projects Operational at end of 1994

Technology	1990 Order (NFFO 1)		1991 Order (NFFO 2)	
	Number of projects generating	Contracted capacity (MW)	Number of projects generating	Contracted capacity (MW)
Hydro	21	10	7	10
Landfill Gas	20	32	26	46
Municipal Solid Waste Incineration	4	41	2	32
Other Waste Incineration	4	45	1	13
Sewage Gas	7	6	19	27
Wind	8	12	25	52
Totals	**64**	**146**	**80**	**180**

Oil Refineries in the United Kingdom[1]

	Distillation capacity	Reforming capacity	Cracking/conversion capacity[2]
	(Million tonnes per annum at the end of 1994)		
Shell UK Ltd			
Stanlow	12.5	2.2	3.8
Shellhaven	4.3	1.6	1.2
Total (Shell)	16.8	3.8	5.0
Esso Petroleum Co. Ltd			
Fawley	15.0	2.8	4.4
British Petroleum Co. Ltd			
Grangemouth	8.9	1.7	2.4
Llandarcy	–[3]	–[3]	–[3]
Total (BP)	8.9	1.7	2.4
Mobil Oil Co. Ltd			
Coryton	8.8	1.6	3.1
Lindsey Oil Refinery Ltd			
South Killingholme	9.4	1.4	4.1
Texaco Refining Co. Ltd			
Pembroke	9.1	1.8	4.5
Conoco Ltd			
Killingholme	6.6	1.9	5.3
Gulf Oil Refining Ltd			
Milford Haven	5.6	1.0	1.6
Elf Oil Ltd/Murco Petroleum Ltd			
Milford Haven	5.5	0.8	1.7
Phillips-Imperial Petroleum Ltd			
North Tees	5.0	–	–
Carless Solvents Ltd			
Harwich	0.6	–	–
Eastham Refinery Ltd			
Eastham	1.1	–	–
Nynas UK AB			
Dundee (Camperdown)	0.5	–	–
Total all refineries	**92.9**	**16.8**	**32.1**

[1] The rated design capacity per stream day multiplied by the average number of days in operation per year.
[2] Includes catalytic, hydro and thermal cracking (including visbreaking).
[3] Crude oil distillation unit at Llandarcy closed January 1986; stand alone vacuum distillation capacity continues in use.

Coal Mines Ownership

Purchaser	Coal mining activities acquired		
	Collieries	**Opencast**	
		Disposal Points	**Operating Sites**

RJB Mining PLC	RJB Mining (UK) Ltd
	
	North
	Kellingly
	Maltby
	Point of Ayr
	Prince of Wales
	Selby Complex:
	    North Selby
	    Riccall
	    Stillingfleet
	    Wistow
	    Whitemoor
	Thorne
	
	Midlands
	Asfordby
	Bilsthorpe
	Daw Mill
	Harworth
	Thoresby
	Welbeck
	
	North East
	Ellington
		North
		Anglers
		Broughton Lodge
		Caroline Staithe
		Oxcroft
		Wardley
		
		Midlands
		Bennerley
		Coalfield North
		Chatterley Valley
		Denby
		Lounge
		Mid Cannock
		Nadins
		
		North East
		Butterwell
		Plenmeller
		Widdrington
			North
			Arkwright
			Broughton Lodge
			Long Row
			Rockingham
			Rye Hill
			
			Midlands
			Bleak House
			Kirk Revised
			Nadins/High Cross
			Rainge
			Smotherfly
			
			North East
			Colliersdean
			Linton Lane
			Plenmeller
			Stobswood

Mining (Scotland) Limited	The Scottish Coal Company Limited Longannet	Blindwells Dalquhandy Damside Killoch Knockshinnock Lambhill Rosslynlee Westfield	Airdsgreen Blindwells Ext. Chalmerston Dalquhandy Damside Lambhill Piper Hill Rosslynlee Westfield Link
Celtic Group Holdings Limited	Celtic Energy Limited	Coedbach Cwmbargoed Gwaun Cae Gurwen Llanilid West Onllwyn	Derlwyn East Pit Extension Ffos Las Gilfach Iago Great White Tip Helid Colliery Kays and Kears Llanilid West Nant Helen
Goitre Tower Anthracite Limited	Tower Colliery Limited Tower		

Lease/License Collieries

Operator	Lease/License Collieries
RJB Mining PLC	Calverton Clipstone Rossington
Coal Investments plc	Hem Heath Coventry Markham Main Silverdale
Betws Anthracite Ltd (MBO)	Betws
Hatfield MEBO	Hatfield

APPENDIX 4 DECOMMISSIONING AND RADIOACTIVE WASTE ISSUES

NIREX

A4.1 As part of the Review of Radioactive Waste Management Policy being conducted by the Department of the Environment, an assessment was made of the economic implications of delaying development of a deep repository for Intermediate Level Waste (ILW). This assessment was summarised in Annex C of the Consultation Paper issued by DoE in August 1994. It was carried out by economists at DTI, DoE, and HM Treasury, using data supplied by UK Nirex Ltd, the nuclear industry, and the Ministry of Defence. This data was also subject to independent review by outside consultants.

A4.2 The assessment concentrated on the financial implications of repository delay for UK Nirex and for the organisations with waste to manage and dispose of. The implications of delay on the costs of the repository itself were assessed, as were the costs of a longer period of interim surface storage of waste. Estimates were also made of the possible additional costs of waste packaging and encapsulation, since delay in finalising the repository site and design would also delay the determination and adoption of optimum packaging specifications.

A4.3 Delaying expenditure on the repository would enable the resources to be put to alternative use in the meantime. Future costs were discounted to reflect the benefits from these alternative uses. Without discounting, there would be no benefit from delay against which to set the extra costs of surface storage. Consequently, the discount rate chosen is crucial to the outcome. Results were calculated for a range of discount rates centred on the 6% rate (in real terms) recommended by HM Treasury for comparing alternative time profiles of public expenditure. Responses to the DoE's consultation paper generally advocated the use of discount rates lower than 6% for this purpose.

A4.4 The assessment concluded that:

- there are very few circumstances where repository delay of 25 years would yield a net benefit in discounted terms;

- for a 50 year delay, the range of outcomes at 6% discount rate could yield a net benefit or net cost, although the balance of probabilities suggests a net benefit equivalent to about 9% of the discounted cost of the early repository alone or about 1.5% of the discounted total cost of waste management and early disposal;

- at 8% discount rate, the probability of a 50 year delay producing a net benefit is high, but at 4% discount rate it is very low.

A4.5 It was recognized that the results of the assessment were subject to very considerable uncertainty on account of the very long timescales under consideration and the complexity of the waste management function and regulatory control thereof.

A4.6 DoE's Consultation Paper also made it clear that a number of other factors not covered in this assessment were relevant to any decision on repository timing. These factors include safety and dose commitments, retrievability, duration of surface storage operations, and compliance with the principles of sustainable development. The Preliminary Conclusions in the Consultation Paper were that, taking the financial and non-financial considerations together, the Government continues to favour a policy of disposal rather than indefinite storage, that Nirex should continue the programme of site investigation but that no fixed deadline should be set for repository completion.

British Nuclear Fuels plc (BNF)

A4.7 BNF's Thermal Oxide Reprocessing Plant (THORP), at Sellafield in Cumbria, started operations in March 1994. The company expects that the plant's commissioning programme will take up to two years to complete. Following commissioning of the Head End plant, which comprises the feed pond, sheer cave and fuel dissolvers, work began on the Chemical Separation area of THORP in January 1995.

A4.8 In late Spring the company began building the Sellafield MOX Plant for making mixed oxide fuel, a blend of plutonium recovered from reprocessing and uranium. This plant is expected to become operational in 1997. BNF's New Oxide Fuels Complex for manufacturing fresh fuel began operations at Springfields near Preston in August 1994.

A4.9 BNFL Inc, BNF's subsidiary in the United States, continues to win contracts for restoring redundant facilities and dealing with radioactive material arising from the American defence programme. The company's order book at March 1994 was worth around $28 million, with 26 contracts won out of the 35 applied for.

Scottish Nuclear Ltd

A4.10 Over the last few years, Scottish Nuclear (SNL) had been developing plans to store certain of its spent AGR fuel from Torness and Hunterston power stations in on-site drystores. Towards the end of 1994, however, BNF offered to SNL revised terms for all aspects of the fuel cycle. These proposals offered commercial attractions to SNL. Contracts have now been signed between SNL and BNF covering the supply of new fuel and the reprocessing or storage of all future SNL spent fuel arisings.

APPENDIX 5   PUBLICATIONS

The following is a selected list of publications produced by Government and other organisations. HMSO publications are available from HMSO Bookshops or:

HMSO Publications Centre (Mail and telephone orders only)
PO Box 276
London SW8 5DT
Tel: 0171 873 9090 (Orders)
Tel: 0171 873 0011 (Enquiries)

Non-HMSO publications are available from the responsible organisation direct. If you have any problems in obtaining these publications please contact:

DTI Library and Information Centre
Department of Trade and Industry
Room 2.LG.4
1 Palace Street
London SW1E 5HE
Tel: 0171 238 3042

This list is correct as at March 1995.

Principal relevant Government publications

1994 Forward Look of Government Funded Science, Engineering & Technology.
HMSO, 1994, 167p and 161p, 2 volumes not sold separately, £30.00 per boxed set, ISBN 0-11-430098-4.

Air Quality: Meeting the Challenge. (Free from the Air Quality Division, Department of the Environment. Tel: 0171 276 8312.)

Atomic Energy Authority Bill: To Make Provision for the Transfer of Property, Rights and Liabilities of the United Kingdom Atomic Energy Authority to other Persons; and for Connected Purposes.
HMSO, 1995, 25p, £5.35, ISBN 0-10-306195-9.

Clean Coal Technologies: A Strategy for the Coal R&D Programme. Energy Paper 63. HMSO, 1994, 44p, £12.95, ISBN 0-11-515359-4.

Climate Change: The UK Programme. Cm 2427. HMSO, 1994, 80p, £10.00, ISBN 0-10-124272-7.

Coal Industry Act. HMSO, 1994, 171p, £15.10, ISBN 0-10-542194-4.

Competitiveness: Helping Business to Win. Cm 2563. HMSO, 1994, 163p, £15.40, ISBN 0-10-125632-9.

Digest of Environmental Protection and Water Statistics 1994. HMSO, 1994, 204p, £15.95, ISBN 0-11-752939-7. (The 1995 edition is due out soon.)

Digest of UK Energy Statistics 1994. HMSO, 1994, 205p, £19.95, ISBN 0-11-515351-9.

Electricity Supply Regulations 1988, SI Number 1988/1057. HMSO, 1988, £4.70, ISBN 0-11-087057-3. Amended by SI Number 1994/533, HMSO, 1994, £1.10, ISBN 0-11-043533-8.

Energy Projections for the UK: Energy Use and Energy-Related Emissions of Carbon Dioxide in the UK, 1995-2020. Energy Paper 65. HMSO, 1995, 162p, £15.95, ISBN 0-11-515365-9.

Energy Related Carbon Emissions in Possible Future Scenarios for the United Kingdom. Energy Paper 59. HMSO, 1992, 52p, £9.95, ISBN 0-11-414157-6.

Energy Technologies for the UK: An Appraisal of UK Energy Research, Development, Demonstration & Dissemination. Energy Paper 61. HMSO, 1994, 69p, £15.95, ISBN 0-11-515386-1. (The detailed information assembled during the appraisal is available – see under Energy Technology Support Unit.)

Energy Trends. (Available from the DTI on annual subscription only. Tel: 0171 238 3606.)

Gas Act Licenses: Memorandum. DTI, March 1995. (Free from DTI, Gas Competition Unit. Tel: 0171 238 3413.)

Gas Bill. HMSO, March 1995, 68p, £8.85, ISBN 0-10-306095-2.

Home Energy Conservation Bill. House of Lords Bill 47, 21 March 1995. HMSO, 1995, 4p, £1.10, ISBN 0-10-870475-0.

Industrial Energy Markets: Energy Markets in UK Manufacturing Industry 1973 to 1993. Energy Paper 64. HMSO, 1994, 209p, £32.00, ISBN 0-11-515352-7.

Minerals Planning Guidance Note number 3: Coal Mining and Colliery Spoil Disposal.
HMSO, 1994, 42p, £6.95, ISBN 0-11-752958-0.

New and Renewable Energy: Future Prospects in the UK. Energy Paper 62.
HMSO, 1994, 113p, £16.95, ISBN 0-11-515384-5. (The detailed information assembled during the assessment is available - see under Energy Technology Support Unit.)

Onshore Licensing Regulations. (To be revised later this year. A single licence, to be called a Petroleum Exploration and Development Licence, will replace the existing three licence system.)

PPG 22: Renewable energy annexes on energy from waste combustion, hydro power, wood fuel, anaerobic digestion, landfill gas, and active solar systems. HMSO, 1994, £4.50, 19p,
ISBN 0-11-753023-9.

Review of Radioactive Waste Management Policy: Preliminary Conclusions. A Consultation Document.
Department of the Environment, August 1994.

Royal Commision on Environmental Pollution, Eighteenth Report, October 1994: Transport and the Environment. Cm 2674. HMSO, 1994, 325p, £25.60, ISBN 0-10-1267742-8.

Sustainable Development: The UK Strategy. Cm 2426. HMSO, 1994, 267p, £22.00,
ISBN 0-10-124262-X.

The Electricity (Class Exemptions from the Requirement for a Licence) Order 1994, SI Number 1070.
HMSO, 1994, £1.15, ISBN 0-11-044070-6.

The Energy Report Volume 1: Markets in Transition. HMSO, 1994, 138p, £25.00,
ISBN 0-11-515379-9.

The Energy Report Volume 2: Oil and Gas Resources of the United Kingdom. HMSO, 1994, 153p, £35.00, ISBN 0-11-515380-2.

This Common Inheritance: 3rd Year Report: Britain's Environmental Strategy. Cm 2549.
HMSO, 1994, 195p, £21.00, ISBN 0-10-125492-X.

This Common Inheritance: UK Annual Report 1995. Cm 2822. HMSO, 1995, 191p, £17.50,
ISBN 0-10-128222-2.

Trade and Industry 1995: The Government's Expenditure Plans 1995-96 to 1997-98. CM 2804. HMSO, 1995, 107p, £16.80, ISBN 0-10-128042-4.

Transport Statistics Great Britain 1994. HMSO, 1994, 233p, £26.00, ISBN 0-11-551633-6.

Other Relevant Publications

ENERGY EFFICIENCY OFFICE, DEPARTMENT OF THE ENVIRONMENT
(2 Marsham Street, London SW1P 3EB)

A full list of publications produced by the Energy Efficiency Office is available from their enquiry point, tel: 0171 276 6200. (A list of Best Practice Programme publications is available from the Energy Technology Support Unit – see below.)

ENERGY TECHNOLOGY SUPPORT UNIT
(Harwell, Didcot, Oxon OX11 0RA. Tel: 01235 432450)

An Appraisal of UK Energy Research, Development, Demonstration & Dissemination. ETSU R 83. HMSO, 1994, two Volumes, £100.00, ISBN 0-11-515349-7. (Energy Paper 61 refers.)

An Assessment of Renewable Energy for the UK. ETSU R 82. HMSO, 1994, 308p, £30.00, ISBN 0-11-515348-9. (Energy Paper 62 refers.)

A list of Best Practice Programme publications is also available.

IEA/OECD

Coal Information 1993. HMSO, 1994, 605p, £85.00, ISBN 92-64-14185-5.

Electricity Information 1993. HMSO, 1994, 499p, £44.00, ISBN 92-64-14184-7.

Energy Balances of OECD Countries. HMSO, 1994, 216p, £29.00, ISBN 92-64-04041-2.

Energy Policies of IEA Countries, 1993 Review. HMSO, 1994, 608p, £79.00, ISBN 92-64-14199-5.

Energy Prices and Taxes, third quarter 1994. HMSO, 1995, 431p, £40.00, ISBN 92-64-14343-2.

Energy Statistics of OECD Countries 1991-1992. HMSO, 1994, 230p, £38.00, ISBN 92-64-04040-4

Energy Statistics and Balances of Non-OECD Countries 1991-1992. HMSO, 1994, 485p, £67.00, ISBN 92-64-04177-X.

Oil and Gas Information 1993. HMSO, 1994, 586p, £85.00, ISBN 92-64-04168-0.

World Energy Outlook, 1994 Edition. HMSO, 1994, 305p, £43.00, ISBN 92-64-14074-3.

WORLD ENERGY COUNCIL

Energy for Tomorrow's World. Kogan Page, 1993, £27.50, ISBN 0-74-941117-1.

EUROPEAN COMMUNITY PUBLICATIONS

EUROSTAT: Monthly Energy Statistics, together with any *Rapid Reports* that are published. Available from HMSO on subscription, currently £77.00 per calendar year. Tel: 0171 873 8409. (Single titles available on request from HMSO, PO Box 276, London SW8 5DT. Tel: 0171 873 9090.)

HOUSE OF COMMONS SELECT COMMITTEES

Environment Committee: 4th Report [session 1992-93]: Energy Efficiency in Buildings. HMSO, 1993, House of Commons papers 648-I, £15.60, ISBN 0-10-024153-0.

Government response to the 4th Report from the House of Commons Select Committee on the Environment [session 1992-93] on Energy Efficiency in Buildings. Cm 2453. HMSO, 1994, £5.50, ISBN 0-10-124532-7.

National Audit Office: Report [93-94 session]: Renewable Energy Research, Development and Demonstration Programme. HMSO, January 1994, House of Commons papers 156, £8.15, ISBN 0-10-215694-8.

Public Accounts Committee response: 42nd Report [93-94 session]: Renewable Energy Research, Development and Demonstration Programme. HMSO, July 1994, House of Commons papers 387, £8.95, ISBN 0-10-238794-X.

Trade and Industry Committee: 1st Report [session 1994-95]: The Domestic Gas Market. Volume 1: Report and Proceedings of the Committee. HMSO, 1994, House of Commons papers 23, £8.95, ISBN 0-10-272795-3.

Government response to the 1st Report from the Trade and Industry Committee [session 1994-95] on the Domestic Gas Market. HMSO, 1995, £4.20, ISBN 0-10-229195-0.

Welsh Affairs Committee: 2nd Report [session 1993-94]: Wind Energy. Volume 1: Report and Proceedings of the Committee. HMSO, 1994, House of Commons papers 336, £13.25, ISBN 0-10-020044-3.

Government response to the 2nd Report from the Welsh Affairs Committee [session 1993-94] on Wind Energy. Cm 2694. HMSO, 1994, £2.00, ISBN 0-10-126942-0.

MONOPOLIES AND MERGERS COMMISSION
(New Court, 48 Carey Street, London WC2A 2JT. Tel: 0171 324 1467)

British Gas plc: Volume 1 of reports under the Gas Act 1986 on the conveyance and storage of gas and the fixing of tariffs for the supply of gas by British Gas plc. Cm 2315. HMSO, 1994, 50p, £7.90, ISBN 0-10-123152-0.

Gas: A report on the matter of the existence or possible existence of a monopoly situation in relation to the supply in Great Britain of gas through pipes to persons other than tariff customers. Cm 500. HMSO, 1994, 138p, £10.30, ISBN 0-10-105002-X.

Gas: Volume 1 of reports under the Fair Trading Act 1973 on the supply within Great Britain of Gas through pipes to tariff and non-tariff customers, and the supply within Great Britain of the conveyance or storage of gas by public gas suppliers. Cm 2314. HMSO, 1994, 62p, £8.65, ISBN 0-10-123142-3.

Gas and British Gas plc: Volume 2 of reports under the Gas and Fair Trading Acts. Cm 2316. HMSO, 1994, 496p, £35.00, ISBN 0-10-123162-8.

Gas and British Gas plc: Volume 3 of reports under the Gas and Fair Trading Acts. Cm 2317. HMSO, 1994, 550p, £38.00, ISBN 0-10-123172-5.

OFFICE OF GAS SUPPLY
(130 Wilton Road, London SW1V 1RQ. Tel: 0171 828 0898)

General Reports

Proposed changes to the gas tariff formula: a consultation document by OFGAS. February 1994.

Gas release scheme: entry rules for the 1994/95 programme. April 1994.

Competition and choice in the gas market: a joint consultation document by OFGAS and DTI. May 1994.

Competition in the non-domestic gas market: a consultation document. May 1994.

Pricing transportation and storage: a summary paper. June 1994.

Proposed price controls on transportation and storage: a consultation document. June 1994.

Price controls on gas transportation and storage: the Director General's decisions. August 1994.

Regulation of the competitive gas market: the way forward. September 1994.

Pricing methology for gas transportation and storage. October 1994.

Charging for pipeline capacity within a network code: a situation report. October 1994.

Development of the gas market within a daily balancing regime: a discussion document. October 1994.

Separation of British Gas' transportation and storage Business from its trading businesses: a consultation document. October 1994.

The efficient use of gas: the role of OFGAS. November 1994.

Demand-side management, a survey of US experience prepared for OFGAS by London Economics. December 1994.

Maximum price for the resale of gas: a guide for landlords and tenants. January 1995.

Maximum charges for the resale of gas: a guide for industrial suppliers and their customers.
January 1995.

Development of the gas market within a daily balancing regime: a summary of responses.
January 1995.

The choice of areas for the initial extension of further competition into the domestic gas market. January 1995.

Separation of British Gas' Transportation and Storage Business from its Trading Businesses – The Director General's Decision. February 1995.

Competitive Market Review. March 1995.

The Annual Report 1993. HMSO, 1994, £8.25, ISBN 0-10-220394-6. (The 1994 Annual Report is due out soon.)

OFFICE OF ELECTRICITY REGULATION
(Hagley House, 83-85 Hagley Road, Edgbaston, Birmingham B16 8QG. Tel: 0121 456 2100)

General

Protecting electricity customers. June 1994.

OFFER *Annual Report 1993.* HMSO, 1994, £12.50, ISBN 0-10-235294-1. (The 1994 Annual Report is due out soon.)

Annual Report 1993: a summary of the main points. June 1994.

Consumer Protection

Maximum resale price for electricity 1994/95. 23 March 1994. Amended 1 October 1994. Amended 1 December 1994. Amended 9 January 1995.

Report on customer services 1993/94. September 1994. Published annually.

Energy Efficiency

Energy Efficiency: standards of performance. March 1994.

Select Committee for the Environment: Energy efficiency in buildings. Response from the DGES. February 1994.

Recommendations on the standards of performance in energy efficiency for the Regional Electricity Companies. A report by the Energy Savings Trust to the Office of Electricity Regulation. March 1994.

Energy Efficiency Standards of performance for Scottish Power and Scottish Hydro-Electric. Consultation paper. November 1994.

Regulation and licensing: General

Trafalgar House bid for Northern Electric: a consultation paper. December 1994.

Competition

The 100kW electricity supply market. January 1994.

The competitive electricity market from 1998. January 1995.

Price control

The distribution price control: proposals. August 1994.

The Scottish distribution and supply price controls: proposals. September 1994.

Generation and the Pool

Consultation on pool reform. March 1994.

Decision on a Monopolies and Mergers Commission reference (Generators). February 1994.

Report on trading outside the pool. July 1994.

Submission to the Nuclear Review. October 1994.

Third Renewables Order for England and Wales. November 1994.

First Scottish Renewables Order. November 1994.

Consultation on own-generation limits of the Regional Electricity Companies. 8 December 1994.

Statement by the Director General on Own-Generation Limits of the Regional Electricity Companies. 20 January 1995.

Pool prices and the Undertaking on Pricing given by National Power and Powergen. 26 January 1995.

Technical

Guidance on the requirements to become a registered meter operator. January 1994.

OFFICE OF ELECTRICITY REGULATION, NORTHERN IRELAND
(Brookmount Buildings, 42 Fountain Street, Belfast BT1 5EE. Tel: 01232 311575)

Annual Report 1994. HMSO, 1995, 36p, ISBN 0-337-09417-9.

Wholesale Electricity Competition in Northern Ireland. Report by the Director General of Electricity Supply Northern Ireland. December 1993.

Northern Ireland and Great Britain electricity prices. A survey. June 1994.

Competition in the Northern Ireland Electricity Market: Proposals for the Operation of a Wholesale Generation Market. December 1994.

THE NATIONAL GRID COMPANY PLC
(National Grid House, Kirby Corner Road, Coventry CV4 8JY. Tel: 01203 537777)

Grid Code. Updated as revised. Current revision January 1995. £35.00 plus servicing and postage. For further details telephone: 01734 363252.

Seven Year Statement. Published annually and updated quarterly. £50.00. For further details telephone: 01203 423065.

Statement of Use of System Charges. Published annually. Free.

EUROPEAN COMMUNITY LEGISLATION

Council Directive of 17 September 1990 on the procurement procedures of entities operating in the water, energy, transport and telecommunications sectors. (90/531/EEC). More commonly known as the "Utilities" Directive.

Council Directive of 23 March 1993 relating to the sulphur content of certain liquid fuels. (93/12/EEC).

European Communities Directive on the limitation of emissions of certain pollutants into the air from large combustion plants. (88/609/EEC). More commonly known as the Large Combustion Plant Directive.

Proposal for a Council Directive introducing a tax on carbon dioxide emissions and energy. COM(92)226.

EC Green Paper of 11 January 1995 on a Community Energy Policy. COM(94)659.

EC proposals of 19 January 1994 for two Council Decisions for the development of TENS in the electricity and gas sectors. COM(93)685.

Amended proposals for Directives concerning common rules for the internal market in electricity and natural gas. COM(93)643. HMSO, 1994, £40.20, ISBN 9-27-762653-4.

EU Licensing Directive 94/22 of 30 June 1994 extending the single market to the oil and gas sector by requiring all licences to be offered on a transparent basis. Official Journal L164.

EC Directive 90/377 of 17 July 1990 on price transparency for gas and electricity charged to industrial end users. Official Journal L185.

Energy labelling of freezers and fridges Directive 94/2/EEC contained wiithin Official Journal L45 of 17 February 1994. HMSO, 1994, £5.25, ISBN 0-11-912830-6.

INTERNATIONAL LEGISLATION

UN Framework Convention on Climate Change. Published by UNEP WMO Informaiton Unit, Geneva.

Protocol to the 1979 Convention on Long-Range Transboundary Air Pollution on further reduction of Sulphur emissions. Cm2718. HMSO, 1994, £5.00, ISBN 0-10-127182-4.

Protocol to the 1979 Convention on Long-Range Transboundary Air Pollution concerning the control of emissions of Volatile Organic Compounds or their transboundary fluxes. Cm 1970. HMSO, 1992, £6.25, ISBN 0-10-119702-0.

1988 Protocol under the Convention on Long-Range Transboundary Air Pollution on the control of emissions of Nitrogen Oxides (commonly known as the Sofia Protocol).

Energy Charter Treaty. Signed 17 December 1994. (Scheduled to be published by HMSO later this year. Further information can be obtained from the DTI International Energy Unit. Tel: 0171 215 4413).

Official Journal of the European Communities L33: Council decision of February 1993 concerning the conclusion of the amendment to the Montreal Protocol on substances that delete the ozone layer. HMSO, 1994, 28p, £5.25, ISBN 0-11-912797-0.

Convention on Nuclear Safety. Published by the IAEA, Vienna. HMSO, 110p, £26.00, ISBN 92-0-102294-8.

Misc.

BP Statistical Review of World Energy. June 1994. Available from BP, Educational Service. Tel: 01202 669940.

Sparking off Efficiency: How the Industry Markets Energy Conservation in Buildings. December 1994. Available from the Association for the Conservation of Energy. Tel: 0171 935 1495.

APPENDIX 6 GLOSSARY OF TERMS

AEA	Atomic Energy Authority
AGR	Advanced Gas-cooled Reactor
BC	British Coal
bcm	billion cubic metres (of gas)
BG	British Gas
BNF	British Nuclear Fuels plc
BP	British Petroleum Company Plc
CCGT	Combined Cycle Gas Turbine
CEE	countries of Central and Eastern Europe
CENELEC	European standards making body
CHP	Combined Heat and Power
CO$_2$	Carbon dioxide
CRINE	Cost Reduction Initiative for the New Era
DGES	Director General of Electricity Supply
DGGS	Director General of Gas Supply
DoE	Department of the Environment
DSB	Demand Side Bidding
DTI	Department of Trade and Industry
ERM	Exchange Rate Mechanism
ESI	Electricity Supply Industry
EU	European Union
FCO	Foreign and Commonwealth Office
FGD	Flue Gas Desulphurisation equipment
FPSO	Floating Production Storage and Off Loading System
FSU	countries of the Former Soviet Union
GDP	Gross Domestic Product
GJ	Gigajoule (=1 thousand million joules)
GW	Gigawatt (= 1 thousand Megawatts)
GWh	Gigawatt hour
HEES	Home Energy Efficiency Scheme
HMIP	Her Majesty's Inspectorate of Pollution
IEA	International Energy Agency
IPCC	Intergovernmental Panel on Climate Change
IPP	Independent Power Producer
JET	Joint European Torus project
JNCC	Joint Nature Conservation Committee
JOULE	EU non-nuclear research and development programme
kW	kilowatt (= 1000 Watts)
kWh	kilowatt hour
LCPD	EC Large Combustion Plant Directive
LEAC	Local Energy Advice Centre
LNG	Liquified Natural Gas
MMC	Monopolies and Mergers Commission
MOX	Mixed Oxide Fuel
MW	Megawatt (= 1000 kilowatts)
NAO	National Audit Office
NE	Nuclear Electric
NFFO	Non-Fossil Fuel Obligation
NGC	National Grid Company
NGLs	Natural Gas Liquids
NIE	Northern Ireland Electricity plc
NOx	Oxides of Nitrogen
NP	National Power
OECD	Organisation for Economic Cooperation and Development
OFFER	Office of Electricity Regulation
OFGAS	Office of Gas Supply
OFT	Office of Fair Trading
OPEC	Organisation of Petroleum Exporting Countries

OSO	Oil and Gas Projects and Supplies Office	**THERMIE**	EU programme for promotion of non-nuclear energy technologies
PCIs	Projects of Common Interest (*EU programme*)		
PEC	Pool Executive Committee	**THORP**	Thermal Oxide Reprocessing Plant
PEDL	Petroleum Exploration and Development Licence	**TPA**	Third Party Access
PES	Public Electricity Supplier	**TWh**	Terawatt hour (TW = 1 thousand GW)
PG	PowerGen		
PHARE	Poland/Hungary Assistance (Regional/Economic) (*EU programme*)	**UKCS**	UK Continental Shelf
		UKOOA	UK Offshore Operators Association
PWR	Pressurised Water Reactor	**UMIS**	Uplift Management Incentive Scheme
QUICS	Qualifying Industrial Consumers Scheme		
		UNECE	United Nations Economic Commission for Europe
RBMK	type of Russian nuclear power station		
		VAT	Value Added Tax
REC	Regional Electricity Company	**VOCs**	Volatile Organic Compounds
RPI	Retail Prices Index	**VVER**	type of Russian nuclear power station
SAP	Standard Assessment Procedure		
		WEC	World Energy Council
SAVE	Specific Actions for Vigorous Energy Efficiency (*EU programme*)		
SCEEMAS	Small Company Environmental and Energy Management Assistance Scheme		
SMP	System Marginal Price		
SNL	Scottish Nuclear Ltd		
SO$_2$	Sulphur dioxide		
SRO	Scottish Renewables Obligation		
TACIS	Technical Assistance to Commonwealth of Independent States (*EU programme*)		
TENs	Trans-European Networks		

APPENDIX 7   MEMBERS OF THE ENERGY ADVISORY PANEL

Sir Martin Holdgate, Chairman. Member of the Royal Commission on Environmental Pollution, and President of the Zoological Society of London.

Dr Mary Archer, Chairman of the National Energy Foundation.

Mr Ronald Campbell OBE, formerly Managing Director of Babcock Energy and Director of the Nuclear Power Group.

Mr Tony Cooper, General Secretary of the Engineers' and Managers' Association, incorporating the Electrical Power Engineer's Association.

Dr Nigel Evans, Managing Director, Caminus Energy Ltd.

Mr David Green MBE, Director of the Combined Heat and Power Association.

Dr Dieter Helm, Director of Oxford Economic Research Associates Ltd, and Fellow of New College, Oxford.

Mr Peter Ibbotson OBE, Departmental Director, J Sainsbury plc.

Professor Alexander Kemp, Department of Economics, University of Aberdeen.

Mr Sam Laidlaw, Managing Director of Amerada Hess Ltd.

Mr Mike Parker, Honorary Fellow of the Science Policy Research Unit (SPRU), the University of Sussex. Formerly Director of Economics, British Coal.

Mr Paul Rich, Managing Director, Allied Steel and Wire Ltd.

Ms Lynda Rouse, Managing Director, Corporate Finance, Barclays de Zoete Wedd Ltd.

Mrs Ann Scully OBE, Vice-Chairman, National Consumers Council, and Chairman, Domestic Coal Consumers' Council.

Dr Anthony White, Head of Corporate Strategy, National Grid Company.

Professor Alan Williams, Department of Fuel and Energy, Leeds University.

INDEX

A
Air quality..........................5.12-5.15

B
British Coal......................8.1-8.3, 8.6
    disposal of properties.................8.13
    privatisation.....................8.8-8.15
    sale of subsidiaries 8.14
Building regulations, amended 3.32

C
Capital equipment, market for..........7.8-7.10
Carbon/Energy tax, proposed
    European Union directive 6.12-6.14
Climate Change Programme. 4.16, 5.18-5.22, 6.29
Coal 8.1-8.17
    mines offered for licensing 8.12
    mining, subsidence...............8.16-8.17
    prices 2.30, 8.6
    regional companies, sale 8.8-8.9
    use in electricity generation 4.22, 8.6
Coal Authority, establishment 8.11
Coal Industry Act.........................8.7
Combined Heat and Power 2.6, 9.64-9.67
Competition.............1.3-1.4, 2.1, 2.38-2.42
Competitiveness
    of UK energy companies 7.2-7.7
    White Paper 3.1
Consumer choice, domestic 2.43-2.44
Convention on Nuclear Safety 6.35
Cost Reduction in the New Era initiative.....4.8

D
Department of the Environment,
    Best Practice Programme 3.24-3.25
Department of Trade and Industry,
    joint consultative document on
    competition in gas market 10.12
DERV, demand trends 12.5
Director-General of Electricity Supply
    proposes new distribution price
    controls 9.26, 9.28, 9.39
Director-General of Gas Supply,
    duties......................10.14-10.15

E
Eastern Europe, UK assistance to 6.34
Electricity
    distribution..................2.5, 9.25-9.28
    generation
        autogenerators 9.62-9.63
        capacity.....................9.7-9.10
        combined cycle gas turbine........4.21
        flue gas desulphurisation..........9.19
        fuels.........................9.4-9.6
        market shares 2.6-2.11
        Orimulsion 9.19
        renewable sources 9.58
    prices.........2.24-2.32, 3.8, 3.11, 9.33-9.36
    regional companies 2.5, 9.1, 9.20,
    ....9.26-9.28, 9.30
    transmission 5.7, 9.20-9.24
        interconnectors 9.23-9.24, 9.40, 9.44-9.45
    voltage harmonisation 6.22
Electricity Pool 9.11-9.12, 9.14-9.18
Electricity supply
    competitive market extended 9.29-9.30
    industry structure 9.1
Energy Charter Treaty................6.23-6.26
Energy costs.........................3.3-3.5
Energy demand
    forecasts......................4.31-4.33
Energy Design Advice Scheme 3.28-3.29
Energy efficiency 3.13-3.36, 6.34
    supplies industry................7.15-7.20
Energy labelling, EC regulations
    on refrigerators and freezers.......3.33, 6.20
Energy Management Assistance Scheme 3.27
Energy prices 2.37, 3.6-3.12
    world 4.4
Energy projections,
    by DTI....................4.9, 4.16-4.25
Energy ratio 3.16, 4.34
Energy technology...................4.26-4.30
European Bank for Reconstruction and
    Development, nuclear safety account ...6.36
European Commission
    draft directive on integrated resource
    planning 6.21

Eco-Management and Audit Scheme ... 3.27
energy efficiency programme
    (SAVE) 6.18-6.20
energy technologies promotion
    programme (THERMIE) 6.16
legal action against national energy
    monopolies.................... 6.11
non-nuclear research and development
    programme (JOULE) 6.16
proposed directives on gas and electricity
    markets..................... 6.9-6.10

European Union
    Energy Green Paper 6.3
    energy labelling regulations for
        refrigerators and freezers 3.33, 6.20
    Large Combustion Plant
        Directive 5.13, 6.15, 12.24
    nuclear safety programmes 6.36
    oil and gas, Licensing Directive 11.9
    proposed directive on carbon/
        energy tax 6.12-6.14

F

Fossil Fuel Levy 2.31
Fuel oil, demand 12.10

G

Gas
    demand trends................. 10.2, 10.4
    prices 2.33-2.36, 4.12, 10.18-10.21
    production 11.2-11.4
    supply...................... 10.1-10.24
        liberalisation of domestic market
        .... 10.7-10.17
            licensing arrangements 10.15
            market shares of suppliers..... 2.18-2.20
    UK-Europe interconnector 10.23
    use for electricity generation
    .... 4.22, 9.4, 10.3-10.4
Gas Bill 10.12-10.14, 10.17
Gas oil, demand....................... 12.9
Gasfields, development............. 11.5-11.6

H

Helping the Earth Begins at Home,
    energy campaign..................... 3.31
Home Energy Conservation Bill............ 3.30
Home Energy Efficiency Scheme 3.34

I

Intergovernmental Negotiating
    Committee on Climate Change 6.30
Intergovernmental Panel on Climatic
    Change 5.17
International Atomic Energy Agency........ 6.37
International Energy Agency...... 4.3, 6.27-6.28

J

Jet fuel, demand 12.8
Joint European Torus 6.38
Joint Nature Conservation Committee,
    coastal regional reports 11.11-11.12

K

Kerosene, delivery pipelines to airports.... 12.12

L

Liberalisation 2.3, 2.38

M

Metering, technology 2.46-2.49
Monopolies and Mergers Commission
    electricity price controls, Scotland,
        reference to 9.39
    reports on gas market 10.7, 10.9

N

National Audit Office, report on
    renewable energy programme 9.59-9.61
National Grid Company 2.5, 9.1, 9.20-9.22
National Power 2.8-2.11
    disposal of capacity............. 9.12-9.13
Natural gas
    exports........................... 10.22
    imports 10.23-10.24
    reserves 11.4
    Scotland-Northern Ireland pipeline 10.5
    Scotland-Republic of Ireland pipeline.. 10.23
    UK-Norway pipeline 10.24

*New and Renewable Energy: Future
   Prospects in the UK* 9.55
Nitrogen oxides, emission control
   initiatives 5.13, 6.32
Non-Fossil Fuel Obligation 4.28, 9.55-9.56
   orders 9.57
Northern Ireland, gas supply 10.5
Northern Ireland
   Electricity plc........... 9.3, 9.40-9.41, 9.45
Nuclear Electric 2.8-2.9, 9.47
Nuclear energy............. 4.24, 9.5, 9.46-9.54
Nuclear fusion, research 6.38
Nuclear power stations
   decommissioning 9.51
   Magnox, closure 9.47
Nuclear Review 4.24, 9.52-9.54

O

Office of Electricity Regulation,
   market share survey 2.14
Office of Fair Trading, review of
   gas market......................... 10.8
Office of Gas Supply, joint consultative
   document on competition in
   gas market........................ 10.12
Offshore exploration, licensing 11.7
Offshore hydrocarbons fields 11.5-11.6
Oil
   prices 4.2, 4.4, 4.7-4.9, 4.13, 12.16-12.21
   production 11.2-11.4
   refining capacity.............. 12.11, 12.15
   reserves 4.5-4.6, 4.8, 11.4
Oil and gas equipment industry 7.11-7.14
Oil and Gas Projects and Supplies Office
   (OSO) 7.11
Oilfields, development 11.5-11.6
Oils, crude, demand 12.14
Onshore Licensing Regulations 11.8-11.9
Organisation of Petroleum Exporting
   Countries (OPEC) 4.4, 4.6-4.7
Orimulsion 9.19, 12.2

P

Petrol, transport demands 12.4
Petroleum fuels
   demand............................ 12.2
   environmental controls 12.23
   exports............................ 12.13
Power stations 9.7-9.10
   flue gas desulphurisation 9.19
PowerGen
   disposal of capacity............... 9.12-9.13
   market share 2.8
Privatisation........................... 2.2
Public Accounts Committee, report on
   renewable energy programme......... 9.61

R

Radioactive waste...................... 9.51
Renewable energy sources 4.28, 9.55-9.61
Royal Commission on Environmental
   Pollution, eighteenth report 12.22

S

Scottish Hydro-Electric 9.2, 9.37, 9.50
Scottish Nuclear....................... 9.50
ScottishPower................... 9.2, 9.37, 9.50
Scottish Renewables Obligation 9.55, 9.57
Sizewell B PWR nuclear power station 9.49
Small Company Environmental and Energy
   Management Assistance Scheme 3.27
Soviet Union, states of former,
   nuclear safety 6.36
Sulphur dioxides, emission
   control initiatives 5.13-5.14, 6.31
Sustainable development 1.2

T

Technology Foresight Programme,
   Energy Panel, report................ 4.30
Trans European Networks 6.5-6.8
Transport fuels
   demand trends 12.3-12.8
   retailing trends 12.6-12.7
Treaty on European Union, draft
   Energy Chapter..................... 6.2

U

UK Continental Shelf
- development 11.3-11.4
- exploration
    - environmentally sensitive areas 11.10
    - licensing 11.7

United Nations
- Nitrogen Protocol 5.13
- Second Sulphur Protocol 5.14, 6.31

United Nations Economic Commission for Europe
- Committee on Energy, Energy Efficiency 2000 Project 6.33
- Convention on Long-Range Transboundary Air Pollution 5.12, 6.31-6.32

United Nations Framework Convention on Climate Change 6.29-6.30

United Nations Intergovernmental Negotiating Committee on Climate Change 6.30

Uplift Management Incentive Scheme 9.18

V

Value-added tax
- extension to domestic electricity bills 9.32
- extension to domestic gas supplies 10.4, 10.20

Volatile Organic Compounds Protocol 5.12, 6.31

W

Wasting Energy Costs the Earth
- campaign, launch of 3.31

World Energy Council 4.3

World markets, competitiveness of
- UK companies 7.2-7.7